30-SECOND
ELEMENTS

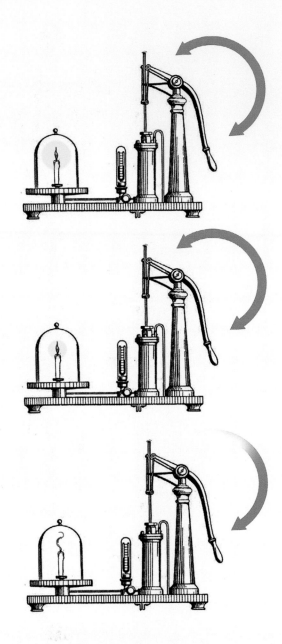

30- SECOND
ELEMENTS

The 50 most significant
elements, each explained
in half a minute

Editor
Eric Scerri

Contributors
Hugh Aldersey-Williams
Philip Ball
Brian Clegg
John Emsley
Mark Leach
Jeffrey Moran
Eric Scerri
Andrea Sella
Philip Stewart

ICON

First published in the UK in 2013 by
Icon Books
Omnibus Business Centre
39–41 North Road
London N7 9DP
email: info@iconbooks.net
www.iconbooks.net

This book was conceived,
designed and produced by

Ivy Press
210 High Street, Lewes,
East Sussex BN7 2NS, UK
www.ivypress.co.uk

Creative Director **Peter Bridgewater**
Publisher **Jason Hook**
Editorial Director **Caroline Earle**
Art Director **Michael Whitehead**
Project Editor **Jamie Pumfrey**
Designer **Ginny Zeal**
Illustrator **Ivan Hissey**
Glossaries Text **Charles Phillips**
Science Editor **Sara Hulse**

ISBN: 978-1-84831-594-5

Printed and bound in China

Colour origination by
Ivy Press Reprographics

10 9 8 7 6 5 4 3 2 1

CONTENTS

INTRODUCTION
Eric Scerri

Interest in the elements and the periodic table has never been greater. Of course, we were all exposed to the periodic table at some point in school chemistry. Everybody remembers the chart that hung on the wall of the chemistry lab or classroom – parts of which we may even have been forced to learn by heart. The chart classifies all the known elements, those most fundamental components that make up the whole earth – and, indeed, the whole universe as far as we know.

But perhaps we didn't appreciate at the time that the periodic table is without doubt one of the most important scientific discoveries ever made and is fundamental to our knowledge of chemistry today. The core idea is deceptively simple – arrange the elements in order of increasing weight of their atoms and every so often we arrive at an element that shows chemical and physical similarities with a previous one. This doesn't just happen occasionally in the periodic table but is true of every single element except the very first few, which, of course, have no earlier counterparts in terms of atomic weight.

Dmitri Mendeleev

Russian chemist Dmitri Mendeleev was the first to propose the periodic table of elements as we know it now, as well as predicting the existence of some undiscovered elements. He is commonly referred to as the father of the peroidic table.

As a result of this behaviour, the linear sequence of elements can be arranged in a two-dimensional grid or 'table', much like a calendar that shows the recurring days of the week for certain dates in a given month. In this way, the tremendous variety among the elements is brought together into a coherent and interrelated form. Of course, these days we use atomic number (number of protons) to order the elements. But although this has solved a technical problem called 'pair reversals', it does not change the periodic table in any major fashion.

At the time Russian chemist Dmitri Mendeleev and a number of others discovered the periodic system, there was no underlying explanation, no apparent reason why the elements should hang together in this way. But the very fact that all the elements could be successfully accommodated into such an elegant system seemed to suggest deeper things to come. Mendeleev was able to use the periodic table to predict the existence of elements that had never been found before. Within about 15 years, three of his best-known predictions came to be The discovery of gallium, scandium, and germanium solidified the notion that the periodic system had latched onto a deep truth about the relationship between the chemical elements.

Around the turn of the 20th century many physicists – including J. J. Thomson, Niels Bohr and Wolfgang Pauli – set about trying to understand what lies behind the periodic system. British physicist Thomson, the discoverer of the electron, was one of the first to suggest that the electrons in an atom are arranged in a specific manner that we now call an electronic configuration. Danish physicist Niels Bohr refined these configurations and provided what remains the gist of the explanation for why elements, in fact, repeat every so often, or to put it another way, why it is that certain elements fall into particular groups or vertical columns in the periodic table. Bohr's answer was that the elements in any vertical column share the same number of outer electrons, building on the emerging notion that chemical reactions are driven by outer electrons.

Next, Austrian physicist Wolfgang Pauli provided a further refinement to the notion of electronic configurations by suggesting that electrons have a hitherto unknown degree of freedom, which became known as 'spin'. To put it in a nutshell, attempts to understand the periodic system contributed enormously to the development of quantum theory and these

Mendeleev's table, 1871
Antoine Lavoisier's 1789 catalogue of elements formed the basis for the modern list, but it was not until Mendeleev's early arrangement in 1869 and final proposal in 1871 that the periodic table was born.

Radioactive research

Pierre and Marie Curie gave their lives in pursuit of discovery, living in poverty to fund their research. Their perilous study of the radioactive elements has since contributed to the understanding of modern nuclear physics.

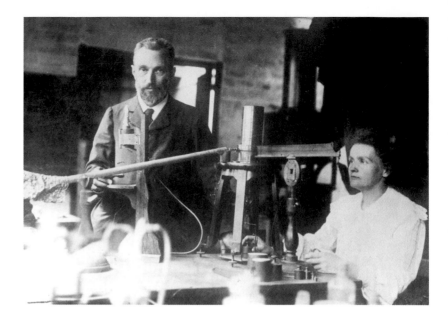

quantum ideas, in turn, provided a theoretical foundation for the periodic system of the elements.

This book will take you on a short and highly digestible tour of 50 of the best known, as well as most intriguing, elements in the periodic table. They range from soft metals, such as sodium, that can be cut with a knife, to the magnificent liquid metal called mercury, to the poisonous gas chlorine that was the world's first chemical weapon.

Each episode is delivered in just 30 seconds and is presented by the world's leading experts on the elements with a proven track record for successfully explaining science to a general audience. You will also learn about the historical figures connected with each of the elements, perhaps as the first to have extracted them – or the first to think they had extracted them because there have been many, many dead ends and failed announcements of the discovery of a good number of the elements.

If condensing an account of each element to 30 seconds were not enough, you will also be provided with a 3-second summary plus a 3-minute version to ponder upon. You will read about the practical applications of the elements, their role in history and how they were first discovered. In some cases, you will read about the disputes that took place before they were eventually discovered or created in particle accelerators. You will learn how each element has its own unique 'personality', even though their atoms ultimately all boil down to different numbers of protons, neutrons and electrons. The chemistry of the elements essentially reduces to the physics of fundamental particles and yet something seems to defy complete reduction. Why else would the elements show such marked individuality while at the same time conforming to the basic physical laws imposed at the quantum mechanical level?

One is reminded of the amazing diversity in the animal kingdom, all of which sprang from one primordial creature. Evolution has also happened in chemistry, providing another facet to the story of the elements. In this case, the earliest form is still with us today. It is the element hydrogen, which still accounts for more than 75 per cent of the mass of the universe. All the other elements have come from hydrogen, directly or mostly indirectly by the fusion of lighter elements that themselves have their origin in hydrogen. Some parts of this astrophysical synthesis took place soon after the Big Bang, while other elements continue to be formed at the centre of stars and galaxies. The heavier elements, by which I mean anything heavier than iron (atomic number 26), are formed under extreme conditions in supernova explosions.

As somebody who has spent a lifetime studying the elements and the periodic table, I am both humbled and proud to have been the consulting editor and to have contributed to this entertaining and informative tour de force on man's new best friends – the elements.

All that glitters ...
Carbon, found in its most common form as graphite, may seem unappealing, but as one of its other many allotropes, as diamond, it is a much sought-after element.

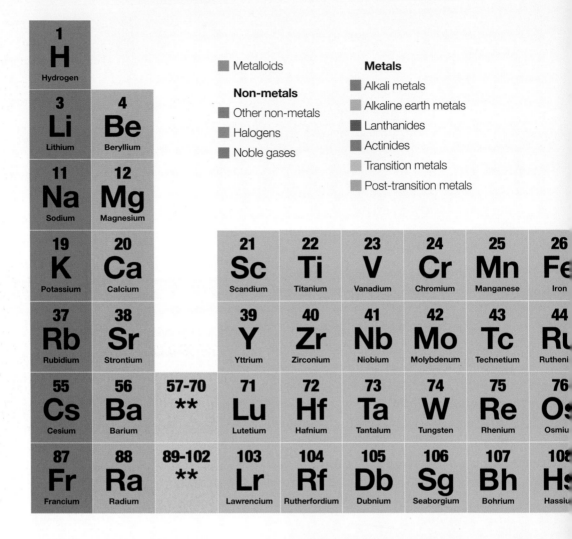

Metalloids

Non-metals
- Other non-metals
- Halogens
- Noble gases

Metals
- Alkali metals
- Alkaline earth metals
- Lanthanides
- Actinides
- Transition metals
- Post-transition metals

The Periodic Table

The Periodic Table is organized by atomic number, electron configuration and recurring chemical properties. The rows of the table are called periods, and the columns are known as the groups. First proposed in 1869 with just 60 elements, the table has expanded to accommodate the 118 elements we now know of.

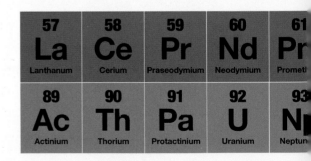

					2 **He** Helium
5 **B** Boron	6 **C** Carbon	7 **N** Nitrogen	8 **O** Oxygen	9 **F** Fluorine	10 **Ne** Neon
13 **Al** Aluminium	14 **Si** Silicon	15 **P** Phosphorus	16 **S** Sulphur	17 **Cl** Chlorine	18 **Ar** Argon

27 **Co** Cobalt	28 **Ni** Nickel	29 **Cu** Copper	30 **Zn** Zinc	31 **Ga** Gallium	32 **Ge** Germanium	33 **As** Arsenic	34 **Se** Selenium	35 **Br** Bromine	36 **Kr** Krypton
45 **Rh** Rhodium	46 **Pd** Palladium	47 **Ag** Silver	48 **Cd** Cadmium	49 **In** Indium	50 **Sn** Tin	51 **Sb** Antimony	52 **Te** Tellurium	53 **I** Iodine	54 **Xe** Xenon
77 **Ir** Iridium	78 **Pt** Platinum	79 **Au** Gold	80 **Hg** Mercury	81 **Tl** Thallium	82 **Pb** Lead	83 **Bi** Bismuth	84 **Po** Polonium	85 **At** Astatine	86 **Rn** Radon
109 **Mt** Meitnerium	110 **Ds** Darmstadtium	111 **Rg** Roentgenium	112 **Cn** Copernicium	113 **Uut** Ununtrium	114 **Fl** Flerovium	115 **Uup** Ununpentium	116 **Lv** Livermorium	117 **Uus** Ununseptium	118 **Uuo** Ununoctium

62 **Sm** Samarium	63 **Eu** Europium	64 **Gd** Gadolinium	65 **Tb** Terbium	66 **Dy** Dysprosium	67 **Ho** Holmium	68 **Er** Erbium	69 **Tm** Thulium	70 **Yb** Ytterbium
94 **Pu** Plutonium	95 **Am** Americium	96 **Cm** Curium	97 **Bk** Berkelium	98 **Cf** Californium	99 **Es** Einsteinium	100 **Fm** Fermium	101 **Md** Mendelevium	102 **No** Nobelium

ALKALI & ALKALINE EARTHS

atom Unit of matter. In an atom, the central nucleus contains positively charged protons and electrically neutral neutrons and is surrounded by negatively charged electrons; in a neutral atom, the number of protons matches the number of electrons.

atomic number The number of protons in an atom's nucleus.

electron shells The electrons surrounding the atomic nucleus are arranged in energy levels – shells or orbitals. The number of electrons in the outer shell or shells defines an atom's chemical properties.

emission spectrum The spectrum of light frequencies emitted by an element when it is heated. Scientists use the emission spectrum to identify the elements combined in a sample; for example, in an alloy used by the steel industry. Astronomers use the spectrum to identify elements present in distant stars and galaxies.

half-life The time taken for half the nuclei in a radioisotope (an unstable isotope) to undergo radioactive decay. The half-life is a measure of how stable a radioisotope is.

hydrous Containing water. A hydrous chemical compound is a 'hydrate'. The opposite, 'anhydrous', describes a compound ('anhydrate') that does not contain water.

ions Electrically charged particles that form when atoms gain or lose electrons.

isotopes Variants of a chemical element with differing numbers of neutrons in the atomic nucleus. All isotopes of an element have the same number of protons in the nucleus. Isotopes can be natural (naturally occurring) or artificial (man-made). Natural isotopes are either stable or unstable. An unstable isotope is said to be radioactive or a radioisotope; the nucleus splits and 'decays', releasing radiation. All artificial isotopes are radioactive.

magic numbers Certain numbers of protons or neutrons in the nucleus make an atom particularly stable, and these are called 'magic numbers'. These are 2, 8, 20, 28, 50, 82 and possibly 114 or 126 and 184. Where there is a magic number of both protons and neutrons, the nucleus is said to be 'doubly magic'.

mass number The total number of protons and neutrons in the nucleus of an atom. Protons and neutrons are together called nucleons and the mass number is sometimes called the nucleon number.

molecule Group of two or more atoms held together by covalent bonds (bonds involving the sharing between atoms of pairs of electrons).

reaction Interaction between two or more molecules resulting in chemical change, typically caused by the movement of electrons between atoms that leads to the breaking or forming of chemical bonds.

reactivity A measure of the tendency of an element or other chemical substance to undergo a chemical reaction. A substance is more reactive if it more readily or quickly tends to react with other substances.

ALKALI AND ALKALINE EARTHS

These elements are in groups 1 and 2 of the periodic table. The alkali metals in group 1 are soft metals, silver in colour, that can be cut with a knife. They all have a single electron in their outer shell and are highly reactive. The alkaline earth metals in group 2 are also silver in colour. They have 2 electrons in their outer shell and, as a result, are less reactive than the alkali metals of group 1. They have higher melting and boiling points than the alkali metals.

Alkali Metals

	Symbol	Atomic Number
Lithium	Li	3
Sodium	Na	11
Potassium	K	19
Rubidium	Rb	37
Cesium	Cs	55
Francium	Fr	87

Alkaline Earth Metals

	Symbol	Atomic Number
Beryllium	Be	4
Magnesium	Mg	12
Calcium	Ca	20
Strontium	Sr	38
Barium	Ba	56
Radium	Ra	88

SODIUM

the 30-second element

This soft, silver-tinted alkali
metal is known for its reactivity. Drop a small
piece into water and it will fizz energetically as
it converts to sodium hydroxide and hydrogen,
giving off plenty of heat. Despite being such a
dramatic element, sodium is named after its
more sedate salt; the word 'sodium' comes
from soda – not a fizzy drink, but sodium
carbonate, an alkaline compound produced
from ashes. It is derived from the Arabic *suda*
('headache') because soda was a popular cure
for headaches; the chemical symbol is short for
natrium, derived from 'natron', the old name for
washing soda or hydrous sodium carbonate.
We come across sodium daily in the yellow
glow of street lamps, produced by the strong
lines in sodium's emission spectrum, but we are
probably most familiar with sodium in common
salt (sodium chloride). Sodium is important for
many living things, including humans. It helps
regulate our blood pressure and builds up the
electrical gradients essential for neurons to fire
in our brains. In modern times, our diets tend to
contain too much salt, resulting in raised blood
pressure and associated health problems.

3-SECOND STATE
Chemical symbol: Na
Atomic number: 11
Named: From 'soda' plus
the metallic element
ending '-ium'

3-MINUTE REACTION
The sodium in salt
originally came from rocks
containing sodium silicate
and sodium carbonate,
dissolved by rivers and
waves crashing over them.
There is no natural isolated
sodium, but the element
occurs widely in minerals,
making it the sixth most
common element in the
earth's crust, around
2.6 per cent by weight. It
is highly reactive due to its
atomic structure, with a
single electron in its outer
shell, which it is more than
enthusiastic to give up.

RELATED ELEMENTS
See also
POTASSIUM (K 19)
page 18

FRANCIUM (Fr 87)
page 20

3-SECOND BIOGRAPHIES
HUMPHRY DAVY
1778–1829
British chemist, first to
isolate sodium

JÖNS JACOB BERZELIUS
1779–1848
Swedish chemist who gave
sodium the Na symbol

30-SECOND TEXT
Brian Clegg

*Sodium regulates
blood pressure and
gives street lamps
their distinctive glow,
but it is most familiar
to us in the form of
sodium chloride (salt).*

POTASSIUM

the 30-second element

Potassium is a soft, silvery, alkali metal that was first isolated by British chemist Humphry Davy in 1807. It is too reactive to be used as a metal, but its salts are important. For centuries, potassium nitrate (saltpetre), potassium carbonate (potash), and potassium aluminium sulphate (alum) have been used in gunpowder, soap making and dyeing, respectively. Today potassium sodium tartrate is used in baking powder, while potassium hydrogen sulphite is added to wines to stop rogue yeasts growing, and potassium benzoate is used as a food preservative. All fertilizers contain potassium, and it is mined on a massive scale – around 35 million tons a year – mainly as the mineral sylvite (potassium chloride). Potassium is used in detergents, glass, pharmaceuticals and medical drips. Around 200 tons per year of potassium metal are produced, and most is converted to potassium superoxide. This is used in submarines and space vehicles to regenerate the oxygen in the air when this has become depleted. Superoxide reacts with CO_2 to form potassium carbonate and oxygen gas. Potassium is an essential element for living things because, along with sodium, it plays a key role in the operation of the nervous system. Potassium-rich foods include peanuts and bananas.

3-MINUTE REACTION
Potassium is a highly reactive alkali metal of group 1 of the periodic table. It exists only as the positively charged potassium ion K^+. Potassium metal dropped into water reacts violently, releasing hydrogen gas, which burns with a lilac flame. Most potassium is the isotope potassium-39, but one atom in 10,000 is potassium-40, which is radioactive – undergoing conversion to argon. This explains why there is 1 per cent of this gas in the earth's atmosphere.

RELATED ELEMENTS
See also
SODIUM (Na 11)
page 16

RUBIDIUM (Rb 37)
page 15

CESIUM (Cs 55)
page 15

3-SECOND BIOGRAPHIES
HUMPHRY DAVY
1778–1829
British chemist who isolated potassium metal for the first time, by means of electrolysis

JUSTUS VON LIEBIG
1803–73
German chemist who in 1840 proved potassium to be an essential element for plants

30-SECOND TEXT
John Emsley

Potassium, which reacts violently with water, is familiar from its use in detergents and glass, while its superoxide plays a key role in recycling air on board submarines.

FRANCIUM

the 30-second element

The existence of element 87 was predicted by Russian chemist Dmitri Mendeleev in 1871 and it was given the provisional name of 'eka-cesium'. A number of scientists searched for the element among non-radioactive sources, but did not find it. The eventual discovery was made by Frenchwoman Marguerite Perey, who had worked as a laboratory assistant to Marie Curie in Paris. Perey became skilful in purifying and manipulating radioactive substances and was asked to examine actinium, element 89 in the periodic table. She was the first to observe the radiation produced by actinium itself rather than its radioactive daughter isotopes; her analysis revealed a new element with a half-life of 21 minutes. When she was later asked to name the element in 1946, she chose francium to honour the country of her birth. Francium was the last natural element to be discovered and it has no commercial applications. However, the fact that it has a very large atomic radius and just one outer-shell electron makes it suitable for atomic physics research. A group in the United States has trapped 300,000 atoms of francium and performed several key experiments.

RELATED ELEMENTS
See also
SODIUM (Na 11)
page 16

POTASSIUM (K 19)
page 18

3-SECOND STATE
Symbol: Fr
Atomic number: 87
Named: After France, the country where the element was discovered

3-MINUTE REACTION
The real interest in francium is an attempt to measure more accurately than before the 'anapole' moment, a new effect predicted by the theory physicists have devised to unify the weak nuclear force with the electromagnetic force. It is called 'anapole', meaning not having to do with any particular kind of pole such as those that appear in the electromagnetic force.

3-SECOND BIOGRAPHIES
FRED ALLISON
1882–1974
American physicist, who was convinced that he had isolated element 87, and published many papers on the subject

HORIA HULUBEI
1896–1972
Romanian atomic/nuclear physicist, who also believed he had isolated element 87

MARGUERITE PEREY
1909–75
French radiochemist, the true discoverer of the last naturally occurring element to be discovered – francium

30-SECOND TEXT
Eric Scerri

Francium's half-life of 21 minutes has made it valuable for research, though estimates suggest there is only 1 oz (30 g) in the earth's crust.

8 February 1834
Born near Tobolsk, Siberia

1855
Teaches at Gymnasium No. 1, Simferopol, Crimea

1859–61
Works in Heidelberg on the capillarity of liquids

1864
Becomes professor at St Petersburg Technological Institute

1865
Chair of chemistry at St Petersburg State University

1865
Publishes dissertation, 'On the Combinations of Water with Alcohol'

1868–70
Writes and publishes *The Principles of Chemistry* in two volumes

1869
Presents "The Dependence between the Properties of the Atomic Weights of the Elements" to The Russian Chemical Society

1882
Married Anna Popova; divorces his first wife, Feozva Leschcheva, a month later

1890
Resigns from St Petersburg State University

1893
Employed at the Department of Weights and Measures

1905
Awarded the Copley Medal from Royal Society

1906
Nominated for Nobel Prize for Chemistry

2 February 1907
Dies of influenza at St Petersburg

DMITRI MENDELEEV

The periodic table of elements

was the brainchild of Russian chemist, academic and civil servant Dmitri Ivanovich Mendeleev. Although he did pioneering work on solutions, gases and the effect of heat on liquids, and helped shape his country's nascent petrochemical industry, Mendeleev will be remembered first and foremost for discovering the periodic table of the elements and using it to predict new elements.

Born in Siberia at the tail end of a long line of siblings (17 – of whom three died before they were christened), he studied at St Petersburg and Heidelberg. In the early 1860s, he gained a professorship at the St Petersburg Technical Institute and soon afterwards took the chair of chemistry at St Petersburg State University. While there, he wrote the definitive textbook on inorganic chemistry, *The Principles of Chemistry* (1868–70, two volumes). In the process of writing, he gradually formulated a table of the 63 then known elements, arranged by atomic weight and valence. He found that they grouped together so coherently he could propose a periodic law, and that the gaps that showed up in the table predicted the existence of elements that had not yet been discovered.

Mendeleev presented his findings to the Russian Chemical Society in 1869, in a paper entitled 'The Dependence between the Properties of the Atomic Weights of the Elements'. Mendeleev claimed to have been unaware of the work done in the same area in the 1860s by Englishman John Newlands and German chemist Lothar Meyer (most notably on periodic behaviour) and there was some controversy when Mendeleev published. Always a colourful character, Mendeleev made what was considered to be a bigamous marriage with Anna Popova in 1882. Russian law at the time stipulated that people had to wait seven years after divorcing before they could remarry.

Despite universal academic plaudits, Mendeleev resigned from the university in 1890 due to his opposition to the government's oppressive treatment of student protests. Three years later he was employed in the Department of Weights and Measures, where he remained until the end of his career, notably working on the standardization of vodka production. In 1905, the Royal Society awarded him the prestigious Copley Medal, and in 1906 he was nominated for the Nobel Prize for Chemistry. This was denied him by the machinations of some scientific rivals, and Mendeleev died a year later. He has his own element, mendelevium, number 101 on the table, first synthesized in 1955.

MAGNESIUM

the 30-second element

Magnesium is a silvery metal that combines strength and lightness, the third most widely used metal worldwide, after iron and aluminium. Adding a few per cent of aluminium to magnesium improves its corrosion resistance and welding qualities: this alloy is used for bicycles, car and aircraft seats, lightweight luggage, lawn mowers and power tools. Magnesium burns with a bright light and magnesium powder was once used for photographic flashbulbs; its most infamous use was in the firebombs dropped during the Second World War. The main magnesium minerals are dolomite and magnesite, forms of magnesium carbonate; 10 million tons are mined per year. Dolomite is used for making the float glass in modern windows. Magnesite is heated to convert it to the oxide and added to fertilizers and cattle-feed supplements, and to make heat-resistant bricks for furnaces. Magnesium is at the heart of the chlorophyll molecule, used by plants to trap carbon dioxide and convert it to carbohydrate. Magnesium is an essential part of our diet, but we store around three years' supply in our body. Foods with high levels are nuts, soyabeans, parsnips, bran and chocolate. Magnesium is abundant in seawater and in the past was extracted from this source.

3-SECOND STATE
Chemical symbol: Mg
Atomic number: 12
Named: After the ancient Greek city of Magnesia

3-MINUTE REACTION
Magnesium is a light metal with a density of 1.7 grams per cubic centimetre (g/cc – much lighter than iron (7.9) and even aluminium (2.7). Once magnesium starts to burn, it is almost impossible to extinguish. It reacts with both oxygen and nitrogen, in the latter case forming magnesium nitride; it is also hard to put out.

RELATED ELEMENTS
See also
CALCIUM (Ca 20)
page 26

STRONTIUM (Sr 38)
page 15

3-SECOND BIOGRAPHIES
JOSEPH BLACK
1728–99
French-Scottish chemist who showed that magnesia (magnesium oxide) was different from lime (calcium oxide) in 1755

HUMPHRY DAVY
1778–1829
British chemist who obtained magnesium metal by electrolysis (in 1808)

30-SECOND TEXT
John Emsley

Magnesium is familiar from its use in traditional flashbulbs and in an aluminium alloy ideal for bicycles. It is most stable as the magnesium ion Mg^{2+}.

CALCIUM

the 30-second element

Calcium is a silvery, fairly soft, white metal, too reactive to be found often as the pure element. It was first isolated in 1808 by British chemist Humphry Davy. After aluminium, it is the most abundant metal in the earth's crust. Over hundreds of millions of years, countless creatures in the oceans, and some on land, have used it to make shells of calcium carbonate; their remains collect on the sea floor and eventually form limestone. Lifted up on to continents, it is slowly dissolved by carbonic acid in rain and carried back to the sea to go through the cycle again, helping to stabilize the level of atmospheric carbon dioxide. Limestone soil is alkaline, and lime-loving plants have a place in nature. Calcium phosphate is a constituent of animal bones and teeth and various physiological processes. Ancient peoples knew the uses of calcium compounds; as early as 4000 BCE, ancient Egyptians heated limestone to prepare lime for use in building. In dry climates, calcium sulphate forms gypsum, from which plaster is still made. In 1823, British engineer Goldsworthy Gurney found that, in a jet of burning hydrogen, lime gives off 'limelight', an early source of stage lighting.

RELATED ELEMENTS
See also
MAGNESIUM (Mg 12)
page 24

FLEROVIUM (Fl 114)
page 150

3-SECOND STATE
Chemical symbol: Ca
Atomic number: 20
Named: From Latin *calx* ('lime')

3-MINUTE REACTION
Calcium is the heaviest element that has a stable isotope (calcium-40) with an equal number of protons and neutrons: 20. This is a 'magic number' (according to the theory of magic nuclear numbers), making calcium one of only four elements that have 'doubly magic' nuclei (a magic number of both protons and neutrons). In fact, calcium is the only element with two doubly magic isotopes: another isotope (calcium-48) has 28 neutrons, another magic nuclear number.

3-SECOND BIOGRAPHIES
HUMPHRY DAVY
1778–1829
British chemist who was first to isolate calcium as well as five other elements. Inventor of the miners' safety lamp

GOLDSWORTHY GURNEY
1793–1875
British engineer and inventor who – besides 'limelight' – also pioneered steam-powered road vehicles and ventilation systems

30-SECOND TEXT
P.J. Stewart

Calcium is a key element in your body, where one of its principal functions is to build and maintain healthy bones and teeth. It is also in your muscles, blood and nervous system.

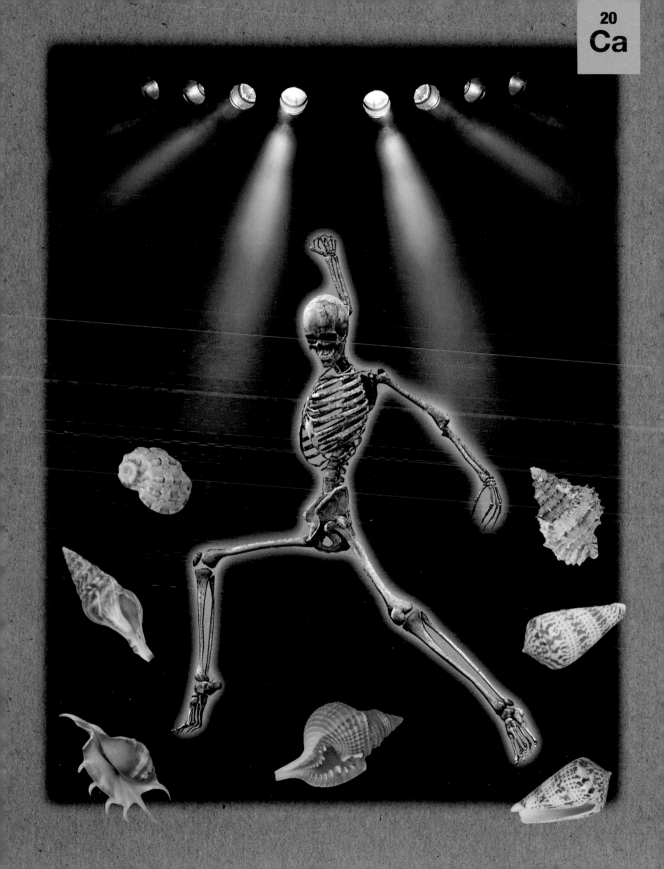

RADIUM

the 30-second element

The alkaline earth metal radium
is the most radioactive substance in nature.
Radium was isolated in 1902 by French-Polish
chemist Marie Curie and husband Pierre from
the waste material left after uranium was
extracted from the mineral pitchblende. This
took months of back-breaking work – the Curies
worked through tons of slag to provide 0.10 g
of radium. After discovering that the new
element produced skin burns when handling it,
the Curies and medical colleagues found that
radiation from the element could destroy
tumours. This 'Curie therapy' was the first
example of radiation-based cancer treatment,
leading to the development of modern
radiotherapy. Radium, with its eerie blue glow,
was seen as a natural source of energy and
incorporated into everything from toothpaste to
hair restorer. It was widely used in luminous
paint, until women workers painting clock dials
began to develop anaemia and cancer. The dial
painters had been licking their brushes to bring
them to a point, ingesting radioactive material;
more than 100 workers died from exposure to
radiation. Marie Curie's own death from aplastic
anaemia was almost certainly a result of
exposure to radiation; even now her notebooks
are kept in lead-lined boxes and handled only
with protective clothing.

3-SECOND STATE
Chemical symbol: Ra
Atomic number: 88
Named: After Latin
radius ('ray')

3-MINUTE REACTION
Radium has four natural
isotopes – variants
featuring differing
numbers of neutrons in
the atom – with atomic
weights ranging from 223
to 228, plus many more
artificial isotopes. The
half-lives of those natural
isotopes vary from
11.4 days to 1600 years,
mostly decaying by
emitting an alpha particle
to produce radon. This
gaseous element is itself
radioactive and can cause
health risks when it builds
up in houses constructed
over minerals with a
concentration of radium.

RELATED ELEMENTS
See also
POLONIUM (Po 84)
page 112

URANIUM (U 92)
page 42

3-SECOND BIOGRAPHIES
PIERRE CURIE
1859–1906
French chemist who
co-discovered radium with
Marie Curie

MARIE CURIE
1867–1934
Polish-French chemist who
discovered and isolated radium

JOHN JACOB LIVINGOOD
1903–86
American chemist who created
radium synthetically, in 1936

30-SECOND TEXT
Brian Clegg

*In addition to its role
in luminescent dials
and hair tonics, this
highly radioactive
element was also
found in tonic drinks
peddled by quack
doctors.*

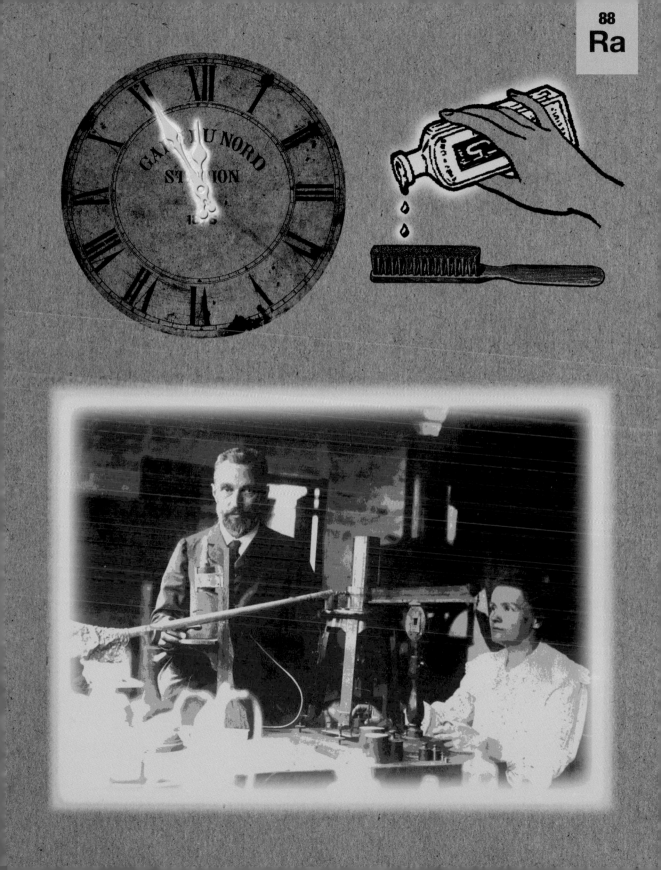

RARE EARTHS

absolute zero −273.15°C (−459.67°F), extrapolated temperature at which atoms have reached so low a kinetic energy that they have almost completely ceased moving.

allotropes Different forms of an element that exist in the same physical state. For example, diamond is an allotrope of carbon.

daughter isotope The product of radioactive decay of an isotope. The original (pre-decay) isotope is called the 'parent isotope'.

ion exchange chromatography Process for separating molecules in a compound on the basis of their charge.

kelvin Temperature scale developed by British chemist William Thomson, Lord Kelvin (1824–1907) in 1848 that takes absolute zero as its starting point. A degree on the scale is equivalent to 1 degree on the Celsius scale, but 0 Kelvin = −273.15°C (−459.67°F). Boiling point for water (100°C/212°F) is therefore 373.15K, and water's freezing point (0°C/32°F) is 273.15K.

nuclear fission The splitting of an atomic nucleus, releasing energy. This process is used in nuclear power stations and some nuclear bombs. Isotopes of uranium and plutonium (uranium-235 and plutonium-239) are usually used as fuel. A neutron fired at a uranium-235 atom splits the nucleus, forming two smaller nuclei, releasing energy and liberating three neutrons.These neutrons hit other uranium-235 nuclei and the process continues in a 'chain reaction'. In a reactor, the reaction has to be controlled to prevent an explosion. The bombs dropped on the Japanese cities of Hiroshima and Nagasaki in the Second World War were both fission bombs, using uranium (Hiroshima) and plutonium (Nagasaki).

nuclear fusion The joining together (fusing) of atomic nuclei to form a single larger nucleus. As with nuclear fission, the process of fusion releases large amounts of energy. Nuclear fusion powers active stars. In our sun, hydrogen nuclei fuse to form helium. The United States produced one bomb powered by nuclear fusion, under the code name 'Ivy Mike', exploded in a test on the Enewetak Atoll, Pacific Ocean, on 1 November 1952.

transuranic element An artificial element with more protons (a higher atomic number) than uranium (which has 92).

trivalent ion An ion with a valency of three.

valency Measure of an atom's combining power, the number of bonds an atom of an element can form with other atoms.

LANTHANIDES (RARE EARTHS) AND ACTINIDES

The lanthanides is a group of elements originally discovered in infrequently found minerals and therefore called 'rare earth elements'; because the elements are found more widely than was once thought, the 'rare earth' label has been dropped from correct usage and the elements are known – as here – as the lanthanides. The lanthanides and actinides occupy the two blocks laid out beneath the periodic table in the conventional layout of the table. Both groups are metallic chemical elements. All lanthanides, aside from promethium, are non-radioactive; actinides are all radioactive.

Lanthanides

	Symbol	Atomic Number
Lanthanum	La	57
Cerium	Ce	58
Praseodymium	Pr	59
Neodymium	Nd	60
Promethium	Pm	61
Samarium	Sm	62
Europium	Eu	63
Gadolinium	Gd	64
Terbium	Tb	65
Dysprosium	Dy	66
Holmium	Ho	67
Erbium	Er	68
Thulium	Tm	69
Ytterbium	Yb	70

Actinides

	Symbol	Atomic Number
Actinium	Ac	89
Thorium	Th	90
Protactinium	Pa	91
Uranium	U	92
Neptunium	Np	93
Plutonium	Pu	94
Americium	Am	95
Curium	Cm	96
Berkelium	Bk	97
Californium	Cf	98
Einsteinium	Es	99
Fermium	Fm	100
Mendelevium	Md	101
Nobelium	No	102

PROMETHIUM

the 30-second element

The discovery of element 61

represented the filling of the final gap within the old limits of the periodic table and was achieved following the discovery of ion-exchange chromatography in the Manhattan Project during the Second World War. The classical methods of separation had failed to discover element 61 because there is simply not enough of the element in the earth's crust. The researchers who synthesized the element in 1945 were not deliberately setting out to form it. The new element, which was eventually called promethium, was identified in the course of attempts to characterize isotopes produced by radiation experiments. Promethium is unusually unstable, and the only one of the 14 lanthanides that is radioactive. Contrary to many published accounts, promethium does occur naturally on earth in minuscule amounts in the mineral apatite and in pitchblende. The isotope Pm-147 is used in nuclear batteries because it is a medium emitter of beta particles that does not produce too much undesirable secondary radiation. Nuclear batteries are expensive, but they tend to have half-lives of as much as 10–20 years, making them much longer-lasting than conventional chemically based batteries. They provide excellent power sources for spacecraft, hearing aids and heart pacemakers.

3-SECOND STATE
Chemical symbol: Pm
Atomic number: 61
Named: From Prometheus, the character in ancient Greek mythology who stole fire from the gods

3-MINUTE REACTION
Until recently, the promethium-based batteries used in space and military applications were large, but a team from the University of Missouri in the United States has now produced batteries the size of a penny and is aiming to reduce the thickness of such batteries to the thickness of a human hair. Such batteries hold up to one million times the charge of a conventional battery.

RELATED ELEMENTS
See also
EUROPIUM (Eu 63)
page 36

GADOLINIUM (Gd 64)
page 38

3-SECOND BIOGRAPHIES
HENRY MOSELEY
1887–1915
English physicist who confirmed that element 61 was missing

JACOB MARINSKY & LAWRENCE GLENDENIN
1918–2005 & 1918–2008
American chemists, co-discoverers of promethium

30-SECOND TEXT
Eric Scerri

Promethium batteries are ideal in cases when it is desirable not to have to change the batteries very often – for example, in heart pacemakers or on board space vehicles.

EUROPIUM

the 30-second element

The rare earth metal europium is a lanthanide, one of the collection of elements that fits between barium and lutetium on the periodic table. The label 'rare earth' is out of date – the rare earths were originally discovered in scarce minerals, but they occur more widely than was first thought. Although a metal, europium is never found naturally in its shiny silvery state because it reacts so easily with the air or water. Europium had three different discoverers. In the late 1880s, British chemist William Crookes found a new spectral line in a mineral that would later be identified as belonging to europium – he was the first to discover the existence of the element. Soon after, French chemist Paul Lecoq de Boisbaudran separated off a material that had the distinctive europium spectral lines, and finally in 1901 another French scientist, Eugene-Anatole Demarçay, isolated a specific europium salt, for which he is usually given the honour of being europium's discoverer. Europium's main practical role is in phosphors (materials that glow when stimulated by electrons or ultraviolet), but it is also excellent at absorbing stray neutrons, and though it has yet to be widely used, it could be valuable in this role to help control nuclear reactors.

3-SECOND STATE
Chemical symbol: Eu
Atomic number: 63
Named: For the continent of Europe

3-MINUTE REACTION
Europium is a versatile doping agent for phosphors. Doping involves adding a small amount of impurity to give a specific colour to the glow produced in phosphors by ultraviolet or electrons. The word 'fluorescent' comes from the mineral fluorite, which has a blue glow thanks to europium salts with valency 2; fluorite is also used in fluorescent tubes, where it is combined with valency 3 europium salts.

RELATED ELEMENTS
See also
PROMETHIUM (Pm 61)
page 34

GADOLINIUM (Gd 64)
page 38

3-SECOND BIOGRAPHIES
WILLIAM CROOKES
1832–1919
British chemist who discovered europium's spectroscopic trace

EUGENE-ANATOLE DEMARÇAY
1852–1904
French chemist who isolated the first europium salt

30-SECOND TEXT
Brian Clegg

Several chemists have had to share credit for discovering europium, the most volatile of the lanthanide elements. Aside from its role as a dopant for phosphors, it is often used in research.

GADOLINIUM

the 30-second element

Gadolinium has an unusual trait shared only by the transition metals iron, cobalt and nickel: ferromagnetism. (This is the mechanism by which certain materials form permanent magnets when placed in a magnetic field; the materials become magnetic and remain so after the external magnetic field is removed.) Gadolinium is a stronger ferromagnet than these three other naturally occurring elements—but only when supercooled to 0 Kelvin (-273.15°C/-459.67°F). When not supercool, gadolinium is ferromagnetic below and paramagnetic above 20°C (68°F); this suggests applications as a magnetic component that can sense hot and cold. (Paramagnetic materials are attracted by a magnetic field but do not retain magnetic properties when the field is removed.) Gadolinium is used in nuclear reactor control rods because it has the highest known capability to absorb neutrons of any natural isotope of any element. Gadolinium gallium garnets and gadolinium yttrium garnets are manufactured for use in microwave applications and in fabrication of various optical components. Found in association with other lanthanides in many minerals, gadolinium occurs in nature in its salts and especially as the oxide, gadolinia, for which it was named.

3-SECOND STATE
Chemical symbol: Gd
Atomic number: 64
Named: For gadolinite, named after Johan Gadolin (1760–1852), who also discovered yttrium

3-MINUTE REACTION
Naturally occurring gadolinium is composed of six stable isotopes and one radioactive isotope, with Gd-158 being the most abundant. Unlike other rare earth elements, metallic gadolinium is relatively stable in dry air. Like most rare earths, gadolinium forms trivalent ions that have fluorescent properties, making gadolinium compound useful as green phosphors in consumer electronics. Gadolinium as a phosphor is also used in other imaging functions such as in X-ray systems.

RELATED ELEMENTS
See also
PROMETHIUM (Pm 61)
page 34

EUROPIUM (Eu 63)
page 36

3-SECOND BIOGRAPHIES
JEAN CHARLES GALISSARD DE MARIGNAC
1817–94
Swiss chemist who studied the rare earth elements, leading to his discovery of ytterbium and co-discovery of gadolinium

PAUL ÉMILE LECOQ DE BOISBAUDRAN
1838–1912
French chemist who isolated gadolinium in 1886

30-SECOND TEXT
Jeffrey Owen Moran

Gadolinium is useful in magnetic resonance imaging (MRI), as well as in X-rays – and it is added to iron, chromium and related alloys to make them easier to work.

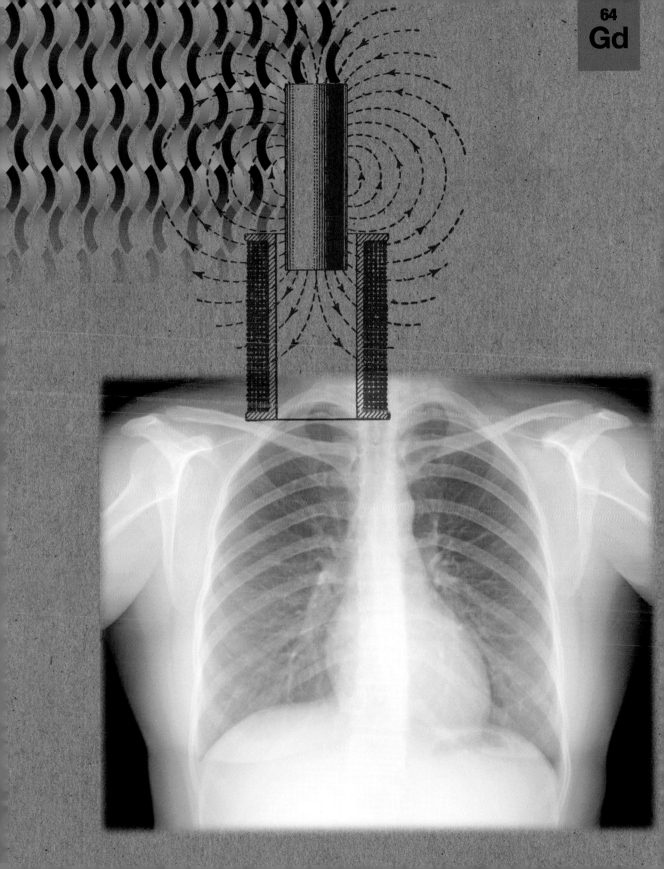

PROTACTINIUM
the 30-second element

Protactinium was one of the few missing elements predicted by Russian chemist Dmitri Mendeleev that was not isolated until well into the 20th century. Mendeleev called it 'eka-tantalum'; he claimed that it should form an oxide with formula R_2O_5, like the elements in the same column of the periodic table, niobium and tantalum. It does – he was right. The first hint of eka-tantalum came from the British chemist-inventor William Crookes, who failed to extract it but identified a new substance that he dubbed 'uranium-X' in uranium ores. In 1913, Polish-American chemist Kazimierz Fajans and his German colleague Oswald Göhring identified an isotope of element 91 and named it brevium in view of its short half-life of 1.17 minutes. Strictly speaking, this represents the first true discovery of the element; however, credit usually goes to the isolation of the longest-lived isotope of a new element, and this was discovered in 1917 by Austrian-Swedish physicist Lise Meitner and German chemist Otto Hahn. Their isotope has a vastly longer half-life of 32,500 years and was named proto-actinium, later shortened to protactinium, meaning the element that forms actinium (element 89) when it loses an alpha particle. The element has virtually no applications due to the fact that it is extremely rare, toxic and highly radioactive.

RELATED ELEMENTS
See also
URANIUM (U 92)
page 42

PLUTONIUM (Pu 94)
page 46

3-SECOND STATE
Chemical symbol: Pa
Atomic number: 91
Named: A slight abbreviation of proto-actinium, meaning 'the element that produces actinium', which it does by alpha decay

3-MINUTE REACTION
Between 1959 and 1961, the UK Atomic Energy Commission succeeded in isolating about 125 g ($4^3/_8$ oz) of protactinium, starting from 60 tons of raw uranium ores. This remains as the largest stash of the element, although it has been somewhat depleted after samples have been sent to scientific establishments around the world.

3-SECOND BIOGRAPHIES
WILLIAM CROOKES
1832–1919
English scientist, journal editor, photographer and inventor

KAZIMIERZ FAJANS
1887–1975
Polish-American radiochemist, discoverer of brevium, a short-lived isotope of element 91

FREDERICK SODDY
1877–1956
English radiochemist who discovered protactinium

30-SECOND TEXT
Eric Scerri

Otto Hahn and Lise Meitner were discoverers of the longest-lived isotope (protactinium-231) of this highly radioactive, silvery-grey metal.

URANIUM

the 30-second element

People used to eat off uranium, and some still do. The element was discovered in the mineral pitchblende in 1789, and in the 19th century it was used to make a bright orange glaze for tableware and a colouring agent for green glass. Orange uranium-ware was still being made in the 1940s, albeit using less radioactive 'depleted uranium', whilst at that same time uranium was being processed by the Manhattan Project into the nuclear bomb that destroyed Hiroshima, Japan, in 1945. The radioactivity of uranium was discovered in 1896 by French scientist Henri Becquerel, who found, while investigating X-rays, that uranium compounds emit a new type of 'ray'. It became clear that the energy leaking from uranium in this way was enormous, and that it was coming from the atomic nuclei. In 1938, German chemists Otto Hahn and Fritz Strassmann in Berlin – together with the Austrian physicist Lise Meitner – found that a uranium nucleus may split in half (undergo fission) when it absorbs a neutron, raising the possibility of a sustained uranium chain reaction that could liberate its nuclear energy more quickly. In a nuclear reactor, the chain reaction is controlled; in a bomb, it becomes a runaway process, releasing the nuclear energy in an explosion.

3-SECOND STATE
Chemical symbol: U
Atomic number: 92
Named: After Uranus, the planet discovered eight years before uranium

3-MINUTE REACTION
Uranium has several isotopes, all of them radioactive. However, only one uranium isotope, denoted U-235, undergoes easier nuclear fission, whereas more than 99 per cent of natural uranium is the barely fissile isotope U-238. So making a bomb demanded that the U-235 be concentrated – a slow process because the isotopes are chemically identical and can't easily be separated. Depleted uranium has some U-235 removed because it is more radioactive (it decays faster).

RELATED ELEMENTS
See also
NEPTUNIUM (Np 93)
page 33

PLUTONIUM (Pu 94)
page 46

3-SECOND BIOGRAPHIES
MARTIN KLAPROTH
(1743–1817)
German chemist who discovered uranium in 1789

HENRI BECQUEREL
(1852–1908)
French physicist who discovered radioactivity ('uranic rays') in uranium

LISE MEITNER
(1878–1968)
Austrian physicist who realized that uranium could undergo nuclear fission

30-SECOND TEXT
Philip Ball

The same element once used to colour earthernware and glassware powered the 'Little Boy' bomb that devastated Hiroshima on 6 August 1945.

19 April 1912
Born in Ishpeming, Michigan

1937
Achieves his doctorate in chemistry from University of California, Berkeley

1937–46
Researches and teaches at UCL Berkeley; appointed professor in 1945

1941
Co-discovers plutonium, with Edwin McMillan, Joseph Kennedy and Arthur Wahl

1942
Joins the Manhattan Project

1944–58
Leads a team that discovers the following elements:
1944: curium and americium
1949: berkelium
1950: californium
1952: einsteinium and fermium
1955: mendelevium
1958: nobelium

1951
Awarded the Nobel Prize for Chemistry

1958–61
Serves as chancellor of the University of California, Berkeley

1961–71
Chairman of the Atomic Energy Commission

1997
Makes history as the first living scientist to have an element named after him – seaborgium

25 February 1999
Dies in Lafayette, California, following a stroke

GLENN T. SEABORG

Glenn T. Seaborg was one of the most important chemists of the 20th century. His discoveries had the greatest impact on the periodic table since Russian chemist Dmitri Mendeleev first proposed the concept in the late 1860s. Seaborg co-discovered and manufactured plutonium and nine other elements among the transuranium elements. He also defined a new set of elements on the periodic table, the actinides.

Seaborg was born in Michigan, 1912, into a family of Swedish immigrants. He studied chemistry at the University of California, Los Angeles, before achieving his doctorate at UCL Berkeley in 1937. He spent most of his career at Berkeley, where he became a professor of chemistry and eventually chancellor. He made his key discoveries using the university's Lawrence Cyclotron.

In 1941, Seaborg co-discovered the element plutonium, with Edwin McMillan, Joseph Kennedy and Arthur Wahl. Plutonium would be used in nuclear reactors and in the atomic bomb dropped on Nagasaki – which Seaborg also helped to develop. During the war, he joined a number of scientists who petitioned US President Truman to stage a public demonstration of the atomic bomb's might, to persuade Japan to surrender. However, their request fell on deaf ears.

In 1940, Edwin McMillan discovered neptunium and over the next 16 years Seaborg and his co-workers at Berkeley discovered the other nine elements in the transuranium sequence, from neptunium (number 93) to nobelium (number 102) – all heavier than uranium.

In 1944, Seaborg inferred that the 14 elements heavier than actinium shared similarities with the element itself and belonged in their own family (the actinides). This sequence included the transuranic elements and showed where they fitted in the table. His theory involved a major redrawing of the periodic table into its current form, with the actinide series running as a strip below the lanthanide series. Seaborg's achievements won him the Nobel Prize for Chemistry in 1951, which he shared with Edwin McMillan.

Seaborg spent years researching nuclear medicine, discovering radioactive isotopes including iodine-131, which enabled his own mother to make a full recovery from thyroid disease. He also advised ten US presidents on atomic energy. Controversially, he was the only scientist to have an element publicly named after him during his lifetime – seaborgium. He remarked that it was 'the greatest honour ever bestowed upon me'.

PLUTONIUM

the 30-second element

3-SECOND STATE
Chemical symbol: Pu
Atomic number: 94
Named: After the
dwarf planet Pluto

3-MINUTE REACTION
Plutonium comes in six allotropes (physical structural variants); it is chemically rich in compounds because it has five oxidation states. This silvery metal, which rapidly oxidizes to a grey sheen, can be used when combined with cobalt and gallium to produce a relatively high-temperature superconductor at around 18.5 Kelvin (−254.65°C/−426.37°F), although its superconducting state is disrupted as the plutonium decays.

You may see the actinide metal plutonium described as the most poisonous substance in existence, but while the element is without doubt poisonous if ingested or inhaled, several natural toxins are more deadly, and in any case, it would be difficult in practice to use plutonium to cause mass poisoning. From its discovery, plutonium has been considered a significant rival to uranium for nuclear energy and the production of nuclear bombs. It is difficult to get a sufficient quantity of plutonium into a critical mass all at the same time, yet if this is achieved 8 kg (18 pounds) of plutonium-239 is enough to produce the damage caused in 1945 by the plutonium bomb dropped at Nagasaki, Japan. Some sources identify uranium as the heaviest of the natural elements, and report that transuranic elements such as plutonium are artificial. However, plutonium does exist in nature: like all the elements heavier than iron, it is created in supernova explosions. We don't see much plutonium on the earth because during the 4.5 billion years of the earth's existence, almost all of our natural plutonium, with the longest half-life at around 80 million years for plutonium-244, has undergone radioactive decay to form uranium.

RELATED ELEMENTS
See also
URANIUM (U 92)
page 42

3-SECOND BIOGRAPHIES
EDWIN MCMILLAN
1907–91
American physicist, first to create a transuranic element (neptunium, in 1940)

GLENN T. SEABORG
1912–99
American chemist who discovered plutonium at Berkeley in 1940

30-SECOND TEXT
Brian Clegg

Created in supernova explosions in distant space, and the power behind the 'Fat Man' bomb detonated over Nagasaki, Japan, on 9 August 1945, plutonium is found on earth in tiny amounts in uranium ores.

HALOGENS & NOBLE GASES

alpha particle Type of particle consisting of two protons and two neutrons, identical to a helium nucleus. It is emitted from an atomic nucleus undergoing alpha decay, a type of radioactive decay.

alpha process One of two classes of nuclear fusion reaction by which stars convert helium into heavier elements. (The other type is the triple-alpha process.)

diatomic molecule A molecule containing just two atoms. These two atoms can be of the same element (for example oxygen, O_2) or two different elements (for example carbon monoxide, CO).

electrolysis Technique that uses an electric current to bring about a chemical reaction. It is used on ionic substances (materials formed when a non-metal reacts with a metal, which contain charged particles or ions). When the current passes through the ionic substance (the electrolyte), negative ions move to the positive electrode (anode) – where they lose electrons and are oxidized – while positive ions move to the negative electrode (cathode), where they gain electrons and are reduced. Electrolysis is used commercially in processes to separate elements from their natural sources such as ores.

fluorinate To combine or treat with fluorine or a fluorine compound.

inert A substance that does not undergo a chemical reaction. The 'noble gases' were at one time known as the inert gases.

nucleon Particle within the nucleus of an atom, either a positively charged proton or an electrically neutral neutron. Each nucleon consists of three quarks.

oxidation Generally a chemical reaction in which atoms react with oxygen, an example being when a metal rusts. More technically oxidation is a reaction in which at least one electron is lost from one of the two substances involved.

oxidation number An artificial, but useful, concept obtained by assuming that an element is bonded ionically and counting the number of electrons gained or lost.

oxidation state Alternate name for oxidation number.

oxide Chemical compound in which oxygen combines with another element.

polymer Chemical compound that contains long chains, typically of carbon atoms. Polymers are created through a process called polymerization.

redox reactions Also called oxidation reduction reactions, chemical reactions in which oxidation and reduction take place.

reduction The opposite of the technical meaning of oxidation – that is, a reaction in which one or more electrons is gained.

substitution reaction Chemical reaction in which one element in a compound (an atom, ion or group of atoms/ions) is replaced by another element.

HALOGENS AND NOBLE GASES

The halogens and noble gases are in groups 17 and 18 of the periodic table. The halogens are non-metallic elements; all have seven electrons in their outer shell. The noble gases all have full outer shells of electrons and so do not readily form compounds. They are called 'noble gases' because they rarely react with other elements – a reference to members of the nobility who traditionally kept themselves aloof from other people in society.

Halogens

	Symbol	Atomic Number
Fluorine	F	9
Chlorine	Cl	17
Bromine	Br	35
Iodine	I	53
Astatine	At	85

Noble Gases

	Symbol	Atomic Number
Helium	He	2
Neon	Ne	10
Argon	Ar	18
Krypton	Kr	36
Xenon	Xe	54
Radon	Rn	86

FLUORINE

the 30-second element

3-SECOND STATE

Chemical symbol: F
Atomic number: 9
Named: From Latin
fluere ('to flow')

3-MINUTE REACTION

Fluorine is the first member of group 17 (the halogens). It will react chemically with all the other elements except helium and neon. Fluorine exists only as the isotope, fluorine-19, which is not radioactive. However, radioactive fluorine-18 can be made and is the basis of positron emission tomography (PET), a medical imaging technique that produces a 3D image of organs and bodily processes: with a half-life of 110 minutes, fluorine-18 decays in a way that allows doctors to monitor the body's vital organs.

Early chemists knew that the mineral fluorspar (calcium fluoride), used in welding metals and etching glass, contained an unknown element, but they could not isolate it. This element was eventually obtained in 1886 by French chemist Henri Moissan, as the pale yellow gas fluorine, by the electrolysis of potassium fluoride in liquid hydrogen fluoride. Fluorine is still produced by this method and is used in the manufacture of products such as Teflon, a fluorinated polymer used for cable insulation, tubing, fabric roofing, plumber's tape, non-stick pans and the wet-weather gear Gore-Tex; another use for Gore-Tex is in artificial veins and arteries. Fluorine is now generally diluted with nitrogen. When polythene containers are treated with this gas, an impenetrable fluorinated layer is formed; these containers make ideal fuel tanks because they are less likely than conventional tanks to rupture in a crash. Fluorine is also used to make uranium hexafluoride, and thereby separate the isotope uranium-235 (the fuel of nuclear reactors). Fluoride (F^-) strengthens bones and teeth by converting the calcium phosphate of which they are made to a harder mineral, fluoroapatite. Some medicinal molecules have a fluorine atom incorporated: these such as the antifungal medicine fluconazole can be highly effective treatments.

RELATED ELEMENTS

See also
CHLORINE (Cl 17)
page 54

BROMINE (Br 35)
page 51

IODINE (I 53)
page 56

3-SECOND BIOGRAPHIES

HENRI MOISSAN
1852–1907
French chemist who first produced fluorine gas in 1886 in Paris

FREDERICK MCKAY
1874–1959
American dentist who proved in the early 1930s that fluoride strengthens teeth

ROY PLUNKETT
1911–94
American chemist who in 1938 discovered Teflon

30-SECOND TEXT

John Emsley

How do chemical elements contribute to cooking eggs? Fluorine (in Teflon) coats your non-stick skillet.

CHLORINE

the 30-second element

3-SECOND STATE
Chemical symbol: Cl
Atomic number: 17
Named: From Greek
khlôros ('pale green')

3-MINUTE REACTION
As a powerful oxidizer, chlorine is widely used as a germ killer. 'Oxidizing' originally referred to a reaction – such as rusting – that involves adding oxygen to a compound, but is now more generally a reaction that removes electrons. When chlorine attacks bacteria, the oxidizing action breaks down the cell membrane, killing the micro-organism. The element is usually added in the form of a compound such as sodium hypochlorite; nevertheless, it is the chlorine that does the work.

A substance that appears to be chlorine was described in the 1630s by Flemish chemist Jan Baptista van Helmont, was spotted in 1774 by German-Swedish chemist Carl Wilhelm Scheele, who called it 'dephlogisticated muriatic acid air', and was identified as an element and given its modern name by British chemist Humphry Davy in 1810. Chlorine is widely available thanks to its presence in seawater. Although we think the sea is salt (sodium chloride) water, it actually contains separate sodium and chloride ions from different sources; salt forms only when the water is evaporated. Chlorine is produced from saline solution by electrolysis: the negatively charged chloride ions are attracted to a positive electrode. Chlorine's disinfectant and antiseptic qualities make it valuable in bleaches and in treating drinking water, in addition to its role in swimming pools. Chlorine's dark side was first unleashed on 22 April 1915, when 6,000 cylinders along the German army's front line were used against Algerian troops of the French army near Ypres. This terrifying weapon was the work of German chemist Fritz Haber. Chlorine burns away the lining of the lungs, leaving victims drowning in the fluid that oozes out.

RELATED ELEMENTS
See also
FLUORINE (F 9)
page 52

IODINE (I 53)
page 56

3-SECOND BIOGRAPHIES
JAN BAPTISTA VAN HELMONT
1579–1644
Flemish chemist who made the first recorded production of chlorine

CARL WILHELM SCHEELE
1742–86
German-Swedish chemist who properly discovered chlorine gas

HUMPHRY DAVY
1778–1829
British chemist who confirmed chlorine as an element and named it

30-SECOND TEXT
Brian Clegg

The distinctive odour of chlorine summons images of swimming pools, yet the element also brings to mind the horrors of chemical warfare in the First World War trenches.

IODINE

the 30-second element

Among the non-metallic elements, iodine is probably the one with which we are most familiar in the home. It lives in the medicine cupboard, where the properties it shares with its fellow halogens – such as the more reactive chlorine – make it ideal for dabbing on cuts and grazes. Canadian poet and singer Leonard Cohen even wrote a song called 'Iodine', which refers to the compassionate sting of its chemical disinfectant action. The discovery of the element is one of the happiest accidents in chemistry. In 1811, French chemist Bernard Courtois used seaweed as a raw material at the family saltpetre works in Paris and noticed a bright violet vapour in the reaction vessel; the vapour condensed to form shiny black crystals. His compatriot Joseph-Louis Gay-Lussac confirmed that this was a new element, and suggested the name iodine. Unfortunately, Courtois was ruined in his efforts to profit from his discovery as his business failed to return on his investment and he stuggled financially for the rest of his life. Iodine's medical applications were quickly realized. It began to be used to treat goitre, a deficiency of the thyroid gland. The knowledge that iodine is relatively abundant in seawater and marine plants explained why sea sponges had proved to be an effective folk remedy for goitre before.

3-SECOND STATE
Chemical symbol: I
Atomic number: 53
Named: From Greek
iodes ('violet')

3-MINUTE REACTION
Because iodine atoms are bulky and form relatively weak single bonds, they are easily detached from certain molecules. This makes the element useful in substitution reactions in the organic chemistry used to make pharmaceuticals. An iodine atom bonded to an intermediate product in a chemical synthesis can be removed and replaced by a complex organic group with particular biomedical functions.

RELATED ELEMENTS
See also
FLUORINE (F 9)
page 52

CHLORINE (Cl 17)
page 54

BROMINE (Br 35)
page 51

3-SECOND BIOGRAPHIES
HUMPHRY DAVY
1778–1829
British chemist mistakenly credited with discovery of iodine

LOUIS DAGUERRE
1787–1851
French scientist who made photographs using iodine-silver reaction

BASIL HETZEL
1922–
Australian campaigner against Third World iodine deficiency

30-SECOND TEXT
Hugh Aldersey-Williams

Commonly found in marine plants, iodine is used in the chemical industry as well as in disinfectants.

2 October 1852
Born in Glasgow,
Scotland

1866–70
Studies at the University
of Glasgow

1870–72
Completes his doctorate
at Tübingen, Germany

1872–80
Researches at Anderson
College, Glasgow, and the
University of Glasgow

1880
Appointed to the chair of
Chemistry at University
College, Bristol, England

1881
Marries Margaret
Buchanan, with whom
he has two children

1887
Becomes professor of
general chemistry at
University College,
London

1888
Elected as a fellow of the
Royal Society

1894
Discovers the new
element argon

1895
Becomes the first
chemist to isolate helium

1898
Discovers krypton, neon
and xenon

1902
Knighted, becoming
Sir William Ramsay

1903
Ramsay and Frederick
Soddy isolate radon
from radium

1904
Wins the Nobel Prize
for Chemistry

1913
Retires and pursues
interests in music, travel
and poetry

23 July 23 1916
Dies in Buckinghamshire,
England, of nasal cancer

SIR WILLIAM RAMSAY

Among the memorials inside Westminster Abbey, London, you'll find names such as Sir Isaac Newton, Lord Nelson, Jane Austen and ... Sir William Ramsay. This Glasgow-born scientist may not be a household name now, but in the late 19th and early 20th centuries he was a scientific celebrity, and in 1904 he became the first Briton to win the Nobel Prize for Chemistry. The accolade was awarded for his discovery of four gases (argon, krypton, neon and xenon), which – along with helium and radon – formed a new family of elements in the periodic table: the noble or inert gases.

Born in 1852, Ramsay was fascinated with chemistry as a child and studied at the University of Glasgow before completing a doctorate at Tübingen, Germany. Returning to Britain, he held a number of academic posts before being appointed to the prestigious chair of general chemistry at University College, London, which he held from 1887 to 1913. It was during this time that Ramsay made his most important discoveries.

In the mid-1890s, he became intrigued by the physicist Lord Rayleigh's observation that nitrogen isolated from air had a slightly higher density than nitrogen obtained from chemical sources. Rayleigh believed there could be an impurity in the chemical sources, but Ramsay suspected that the anomaly could be caused by the presence of an undiscovered element in air that was mixed with the nitrogen. This marked the beginning of a fruitful collaboration with Rayleigh to test his theory.

Ramsay separated nitrogen from air and then passed it over heated magnesium, which produced a solid substance (magnesium nitride). He was left with around 1 per cent of residual gas that would not react, and on measuring it found it to be denser than nitrogen. The gas produced unique spectral lines and Ramsay knew he had found a new element in air, which he called argon.

Ramsay later isolated helium from the mineral cleveite, discovered the elements krypton, neon and xenon, and in 1910, proved that radon was a noble gas. Ramsay identified a whole new group in the periodic table, the noble gases, and Dmitri Mendeleev, the Russian chemist who discovered the table, eventually accepted his findings. The dapper Scottish chemist had made scientific history.

HELIUM

the 30-second element

3-SECOND STATE

Chemical symbol: He
Atomic number: 2
Named: For Greek *helios*
('sun'), where the element
was first discovered

3-MINUTE REACTION

Helium was one of the first elements to exist, created shortly after the Big Bang as the initial superheated 'soup' of matter cooled to form atoms, but more has been produced since by stars. The primary fuel of stars is hydrogen, which forms helium in nuclear fusion reactions. The helium nucleus, consisting of two protons and two neutrons, is also a common product of radioactive decay, where the nuclei are known as alpha particles.

This light noble gas makes up nearly one-quarter of the matter of the universe by mass – yet it was not noticed until 1868, when British astronomer Norman Lockyer found previously unknown spectral lines in the sun. Spectroscopy identifies elements present in a substance from dark lines that appear in a rainbow of light passed through a gas. Each element has a fingerprint of lines that can be used to identify it, even in the remote light from stars. French astronomer Jules Janssen also spotted the unexpected lines, but it was Lockyer who correctly identified this as a new element, which he named helium. In the 1890s, British chemist William Ramsay produced the gas when dissolving the mineral cleveite in acid. Helium is now primarily obtained as a by-product of natural gas extraction. We are familiar with helium in lighter-than-air party balloons. The gas produces a distinctive squeaky voice when inhaled because the speed of sound in helium is significantly higher than in air. Helium has also been used in airships. In addition, its very low boiling point of just above 4 kelvin (−269°C/−452°F) makes it ideal for cooling speciality equipment such as the magnets of MRI scanners.

RELATED ELEMENTS

See also
NEON (Ne 10)
page 64

ARGON (Ar 18)
page 66

KRYPTON (Kr 36)
page 51

3-SECOND BIOGRAPHIES

NORMAN LOCKYER
1836–1920
British astronomer who discovered helium's spectral lines in the sun

WILLIAM RAMSAY
1852–1916
British chemist who isolated helium

30-SECOND TEXT
Brian Clegg

We know helium from the balloons that rise to the ceiling at parties, and helium airships were once the future of aviation. Norman Lockyer discovered the element.

NEON

the 30-second element

During the 1890s, new technology used to liquefy air enabled the British chemist William Ramsay to separate out minor gaseous constituents whose existence had hardly been suspected. He isolated five new chemical elements now known as noble gases, including neon, which constitutes about one part in 60,000 of air. This remarkable achievement earned him the Nobel Prize in Chemistry in 1904. Ramsay confirmed that these gases were, indeed, unique elements by examining the characteristic spectrum of light they produce when excited by an electric discharge. His co-worker, fellow British chemist Morris Travers, was thrilled to note a 'blaze of crimson light' in the case of neon. Neon is so inert that it forms no chemical compounds at all. Yet, despite this lack of reactivity, it is one of the elements most widely known outside the laboratory – because of this signature light. The light is produced when electrons in orbit around the nuclei of neon atoms, having been excited by the electric discharge, return to their normal positions, releasing the energy they have absorbed. In the modern era, 'neon' has become a byword for all kinds of illuminated messages, although neon itself only produces red light. Other inert gases yield other colours when they are similarly placed in an electric discharge.

3-SECOND STATE
Chemical symbol: Ne
Atomic number: 10
Named: From Greek
neon ('new')

3-MINUTE REACTION
One of the lightest and chemically most stable of the elements, neon is the fifth most abundant element in the universe – after hydrogen, helium, oxygen and carbon. It is a product of the alpha process, in which heavy elements are formed by the addition of alpha particles or helium nuclei to lighter atoms. Helium nuclei contain four nucleons (two protons and two neutrons), while carbon, oxygen and neon contain 12, 16 and 20 nucleons respectively.

RELATED ELEMENTS
See also
HELIUM (He 2)
page 62

ARGON (Ar 18)
page 66

XENON (Xe 54)
page 51

3-SECOND BIOGRAPHIES
DMITRI MENDELEEV
1834–1907
Russian chemist who resisted extending the periodic table to accommodate Ramsay's new elements

GEORGES CLAUDE
1870–1960
French innovator of neon signs

BRUCE NAUMAN
1941–
American artist who often works with neon

30-SECOND TEXT
Hugh Aldersey-Williams

Bright neon lights do not obscure William Ramsay, discoverer of neon and other noble gases.

ARGON

the 30-second element

Argon is all around us – there are about 50 trillion tons of it in the atmosphere – yet we didn't catch up with it until the late 19th century. That's because argon doesn't do anything to make itself known. Like the other elements that share its column in the periodic table, it is an inert gas: unreactive, if not downright lazy (the quality for which it's named). Argon does not lose or share any electrons by undergoing chemical reactions; it has a so-called filled shell of electrons. It was discovered in studies of the composition of air, of which it makes up 1 per cent. British chemist Henry Cavendish noticed an inert fraction of air in 1785, but didn't follow it up, and not until 1894 did his compatriots Lord Rayleigh and William Ramsay isolate the inert component from atmospheric nitrogen. Today, three quarters of a million tons of argon are extracted annually from liquefied air, because its very inertness makes it useful. You can fill light bulbs, fluorescent tubes and double-glazed windows with it, or use it as a propellant for aerosols, industrial sprays and even futuristic ion-propulsion spacecraft engines, without worrying that it will react or be toxic.

3-SECOND STATE
Chemical symbol: Ar
Atomic number: 18
Named: From Greek
argos ('lazy, inactive')

3-MINUTE REACTION
It is possible to make argon react with other elements, but only just. In 2000, a team from the University of Helsinki reported that they had reacted argon with hydrogen fluoride in a matrix of solid, frozen argon at a temperature of –246°C (–411°F). The result was argon fluorohydride, chemical formula HArF – but you can barely so much as look at it without it decomposing.

RELATED ELEMENTS
See also
NEON (Ne 10)
page 64

KRYPTON (Kr 36)
page 51

XENON (Xe 54)
page 51

3-SECOND BIOGRAPHIES
HENRY CAVENDISH
1731–1810
British chemist who identified first hint of argon in 1785

JOHN STRUTT
(LORD RAYLEIGH) &
WILLIAM RAMSAY
1842–1919 & 1852–1916
British chemists, co-discoverers of argon in 1894

30-SECOND TEXT
Philip Ball

Argon was difficult to isolate – Lord Rayleigh declared it cost 'a thousand times its weight in gold' – but it has many modern uses and today is produced in vast quantities.

alloy A material that has metallic properties and is composed of two or more chemical elements of which at least one is a metal. Alloys are usually harder than pure metal and more resistant to corrosion. Brass is an alloy made from 70 per cent copper and 30 per cent zinc; bronze is an alloy of 90 per cent copper and 10 per cent tin.

amalgam An alloy of mercury with another metal. Iron does not form an amalgam with mercury but most other metals do. Dental amalgam, combining mercury with silver, tin and other metals, was popular from around 1800 onwards, but is less used today because of health concerns over the use of mercury.

ductile Capable of being pulled out into a wire.

heavy metal One of a group of elements among transition metals, metalloids, lanthanides and actinides that have metallic properties. The term generally refers to those that are heavier than iron and zinc. Examples include mercury, lead and cadmium. They are toxic to humans if ingested.

insulator Material that prevents an electrical charge flowing through it.

isomers Compounds with an identical molecular formula but different structural formulae. While a molecular formula describes the combination of elements in a molecular compound, the structural formula describes how the atoms are fitted together in the molecule. For example, isomers of hydrocarbons have the same number of hydrogen and carbon atoms, but they are connected in different ways.

metastable state A relatively stable state of an atom or molecule, more stable than its most excited states, but less stable than its most stable state.

nuclear isomer A metastable state of an atomic nucleus, in which one or more of its protons or neutrons is excited (has an elevated level of energy).

ore Rock containing a valuable element (typically a metal), for which it is mined.

photon A quantum (bundle) of electromagnetic energy.

quadruple bond Bond between two atoms that involves eight bonding electrons. A single bond involves two electrons, a double bond four and a triple bond six. Quadruple bonds are most commonly made among transition metals such as rhenium and chromium.

salts Ionic compounds formed when an acid undergoes a neutralization reaction with a base.

superheavy element Another name for transuranic elements (see page 32–33), elements with an atomic number greater than 92 (the atomic number of uranium). In some contexts, however, superheavy element refers to elements with an atomic number greater than 100.

TRANSITION METALS

The transition metals are in groups 3–12 of the periodic table. They are mostly dense and hard, and are good conductors of electricity and heat. Their valence electrons (with which they combine with other elements) are in more than one electron shell.

Transition metals

	Symbol	Atomic Number
Scandium	Sc	21
Titanium	Ti	22
Vanadium	V	23
Chromium	Cr	24
Manganese	Mn	25
Iron	Fe	26
Cobalt	Co	27
Nickel	Ni	28

	Symbol	Atomic Number
Copper	Cu	29
Zinc	Zn	30
Yttrium	Y	39
Zirconium	Zr	40
Niobium	Nb	41
Molybdenum	Mo	42
Technetium	Tc	43
Ruthenium	Ru	44
Rhodium	Rh	45
Palladium	Pd	46
Silver	Ag	47
Cadmium	Cd	48
Lutetium	Lu	71
Hafnium	Hf	72
Tantalum	Ta	73
Tungsten	W	74
Rhenium	Re	75
Osmium	Os	76
Iridium	Ir	77
Platinum	Pt	78
Gold	Au	79
Mercury	Hg	80
Lawrencium	Lr	103
Rutherfordium	Rf	104
Dubnium	Db	105
Seaborgium	Sg	106
Bohrium	Bh	107
Hassium	Hs	108
Meitnerium	Mt	109
Darmstadtium	Ds	110
Roentgenium	Rg	111
Copernicium	Cn	112

CHROMIUM

the 30-second element

Chromium is one of the so-called transition metals (like iron, cobalt, nickel and copper). Its compounds form the basis of many traditional artists' pigments and paints; chrome yellow, for example, is pure lead chromate. The colour of rubies and emeralds is due to contamination of otherwise transparent crystalline material with small amounts of chromium oxide. Chromium was discovered in 1798 by French chemist Louis-Nicolas Vauquelin, who ground up precious stones in an effort to explain their colours, and later discovered the element beryllium in the same way. Chromium alloyed with iron makes stainless steel, which does not oxidize or rust. Stainless steel in cutlery might contain as much as 18 per cent chromium, while that for marine use may contain even more. However, it is in the form of chrome plating that we know the element best. This process only became possible on a wide scale with the commercialization of electroplating in the 1920s. 'Chrome' was adopted as a symbol of the consumer society. In 1933, American etiquette expert Emily Post hailed it as the 'answer to the housewife's prayer'. More recently, though, it seems that the familiar thin layer of chrome has come to denote a merely superficial glamour.

3-SECOND STATE
Chemical symbol: Cr
Atomic number: 24
Named: From Greek
chroma ('colour')

3-MINUTE REACTION
Solutions of chromium sulphate have been used since the mid-19th century in the tanning of leather to make it water resistant. The chemistry of the interaction between the inorganic chromium complexes and the organic collagen in the leather is highly involved. Human exposure to these chromate salts can lead to ulcers, and tanneries have released the salts into rivers. Other chromium compounds are still more hazardous, increasing environmental concerns.

RELATED ELEMENTS
See also
IRON (Fe 26)
page 74

COPPER (Cu 29)
page 76

3-SECOND BIOGRAPHIES
COLIN G. FINK
1881–1953
American chemist who perfected chrome plating at Columbia University

HARLEY EARL
1893–1969
American industrial designer, the 'da Vinci of Detroit', responsible for extravagant chrome styling on cars

30-SECOND TEXT
Hugh Aldersey-Williams

Chromium's many oxidation states lead to compounds with a wide range of colours admired by artists, but we are most familiar with this element through the use of chrome plating.

IRON

the 30-second element

As iron ore (oxide) and other minerals, iron makes up about 5 per cent of the earth's crust, and is the fourth most abundant element there. The earth's core is mostly iron, molten in the outer core and solid in the middle; the sloshing of magnetic liquid iron creates the geomagnetic field, which helps protect life from the solar wind. Iron in haemoglobin makes blood red and ferries oxygen. The importance of iron can be judged from the use of the phrase 'Iron Age' to describe a period of human history (beginning in the Middle East in about 1500 BCE): the Hittites, early iron smelters, trampled over Asia Minor, just as the iron-clad Romans later conquered half the world. Swords made in the earlier Bronze Age didn't stand a chance against hard, gleaming steel. Steel is iron mixed with a little carbon, which makes it harder. Because charcoal is used to extract iron from its ore, what you get is inevitably steel instead of pure, softer iron. The best steel requires precise control of carbon content, which became possible in the mid-19th century; only then could engineers build steel bridges without fear that the structures would crack.

3-SECOND STATE
Chemical symbol: Fe
Atomic number: 26
Named: From Anglo-Saxon iren; Fe from Latin ferrum

3-MINUTE REACTION
Of all elements, iron has the most stable nucleus, prone neither to nuclear fusion (merging) or fission (splitting). Crudely put, this stability comes from an ideal balance of constituents. With fewer nuclear particles (protons and neutrons), the nucleus has too much surface, prompting droplet-like mergers; with more protons, there's too much electrical repulsion. So nuclear fusion in stars stops when the constituents are converted to iron.

RELATED ELEMENTS
See also
CHROMIUM (Cr 24)
page 72

NICKEL (Ni 28)
page 71

COPPER (Cu 29)
page 76

3-SECOND BIOGRAPHIES
TOBERN BERGMANN
1735–84
Swedish chemist who established how carbon dictates the properties of steel

HENRY BESSEMER
1813–98
British engineer who invented modern steel-making

30-SECOND TEXT
Philip Ball

Iron colours the surface of Mars, the 'red planet', and gives its name to Ironbridge Gorge in Shropshire, England, where a 30-m (100-foot)-span iron bridge was built in 1779–81.

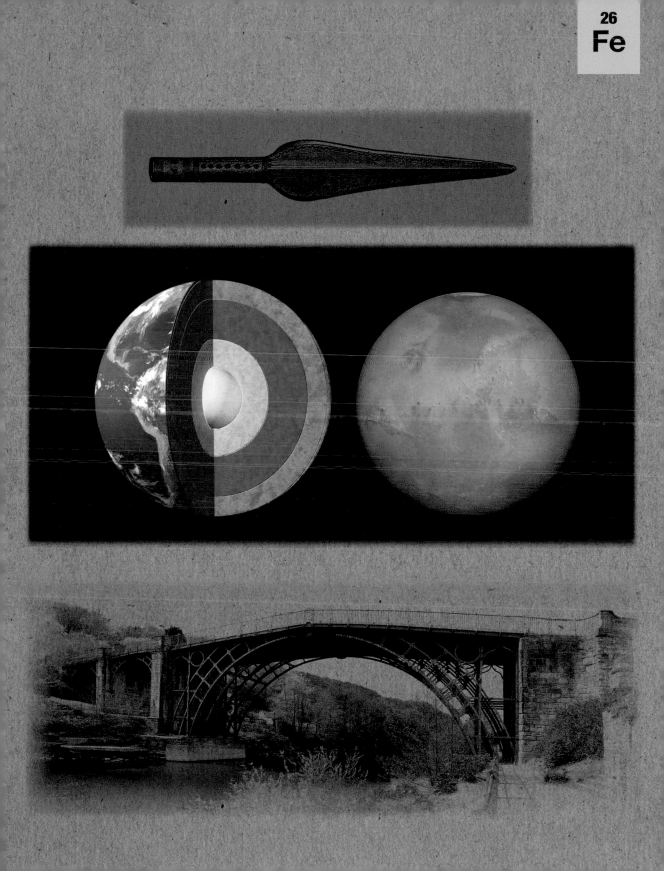

COPPER

the 30-second element

Familiar, reddish-orange copper
is not considered a precious metal, but it is
precious enough to those who strip it from
unwatched buildings. An unusually good
conductor of both heat and electricity, copper
is widely used for sheets, wires, pipes and
fittings. Found naturally in the free state, copper
occurs combined in many minerals, usually in
association with sulphur. It is harder than zinc
but softer than iron, and acquires strength and
structure by mixing with other metals in more
than 1,000 combinations. Combined with
10 per cent tin, it forms the alloy bronze, which
gave its name to an age of human development
three millennia long (roughly 3600 BCE—600 BCE),
when weapons and implements were chiefly
made of copper and bronze. Freshly exposed
copper has street appeal, ages gracefully to an
earthy mahogany and, with weather, becomes
robed – like the Statue of Liberty in New
York City – in a patina of verdigris. Its
compounds, commonly encountered as
copper(II) salts, often produce blue or green
colours in such minerals as turquoise and
malachite. The element is present in minute
amounts in the animal body, and is essential
to normal metabolism.

3-SECOND STATE
Chemical symbol: Cu
Atomic number: 29
Named: From *cypriumaes*,
the latin for 'Cyprus
metal' – Cyprus was
the chief source of
copper in Roman times

3-MINUTE REACTION
Natural copper is a mixture
of two stable isotopes:
Cu-63 (69.17 per cent) and
Cu-65 (30.83 per cent).
Copper has low chemical
reactivity and resists
corrosion, forming a layer
of brown-black copper
oxide, and, eventually, a
green layer of copper
carbonate. Like silver and
gold, copper's fellows in
group 11, its atoms form
relatively weak metallic
bonds, rendering these
metals extremely malleable
and ductile and imparting
exceptionally good thermal
and electrical conductivity.

RELATED ELEMENTS
See also
SILVER (Ag 47)
page 82

TIN (Sn 50)
page 126

GOLD (Au 79)
page 88

3-SECOND BIOGRAPHIES
FRÉDÉRIC BARTHOLDI
1834–1904
French designer responsible
for the Statue of Liberty

WILLIAM A. CLARK, MARCUS
DALY & F. AUGUSTUS HEINZE,
KNOWN AS 'THE COPPER
KINGS OF MONTANA'
Fl. late 19th century
American entrepreneurs
who fought over Montana's
copper-mining industry

30-SECOND TEXT
Jeffrey Owen Moran

*The Statue of Liberty,
erected in 1886, has a
copper exterior and
was originally a copper
colour, but it developed
a green 'patina' as the
copper oxidized in the
years after 1900.*

TECHNETIUM

the 30-second element

Technetium, or element 43, was first synthesized in 1937. The experiments took place in Berkeley, California, but the new element's discovery had to wait until plates of irradiated molybdenum had been sent to Sicily. There, Italian physicist Emilio Segrè, who had recently returned home after working in Berkeley, and a chemist colleague, Carlo Perrier, discovered that element 43 had been created as a result of irradiation – the first artificially synthesized element. Technetium was later found to occur naturally on the earth but only in minuscule amounts. Its rarity is surprising given its relatively low atomic number of 43; the full explanation is complicated, but it is connected with the fact that its isotopes contain odd numbers of protons and also neutrons. Among many other applications, technetium is used in hospitals in medical imaging. This involves one isotope of the element, Tc-99, and a metastable nuclear isomer of this isotope. What makes this isotope especially useful is that it has a half-life of about six hours. This means that over a period of 24 hours about 94 per cent of the technetium isotope decays from the body.

3-SECOND STATE
Chemical symbol: Tc
Atomic number: 43
Named: From Greek
technos ('artificial')

3-MINUTE REACTION
German chemists Ida and Walter Noddack claimed to have discovered element 43 in 1925, calling it 'masurium', and as recently as the turn of the 21st century, two physicists – the Belgian Pieter van Assche and the American John T. Armstrong – were still claiming that the Noddacks had, in fact, isolated element 43 in 1925. This claim has now finally been refuted by a number of other authors.

RELATED ELEMENTS
See also:
MANGANESE (Mn 25)
page 71

RHENIUM (Re 75)
page 86

BOHRIUM (Bh 107)
page 71

3-SECOND BIOGRAPHIES
WALTER NODDACK &
IDA NODDACK
1893–1960 & 1896–1978
German chemists who claimed to have discovered element 43 in Berlin in 1925

CARLO PERRIER & EMILIO
SEGRÈ
1886–1948 & 1905–89
Italian physicist and chemist who were the true co-discoverers of element 43 in Palermo in 1937

30-SECOND TEXT
Eric Scerri

The technetium isotope, Tc-99, is used in around 50 million medical imaging procedures carried out each year and has also been proposed for use in nuclear batteries.

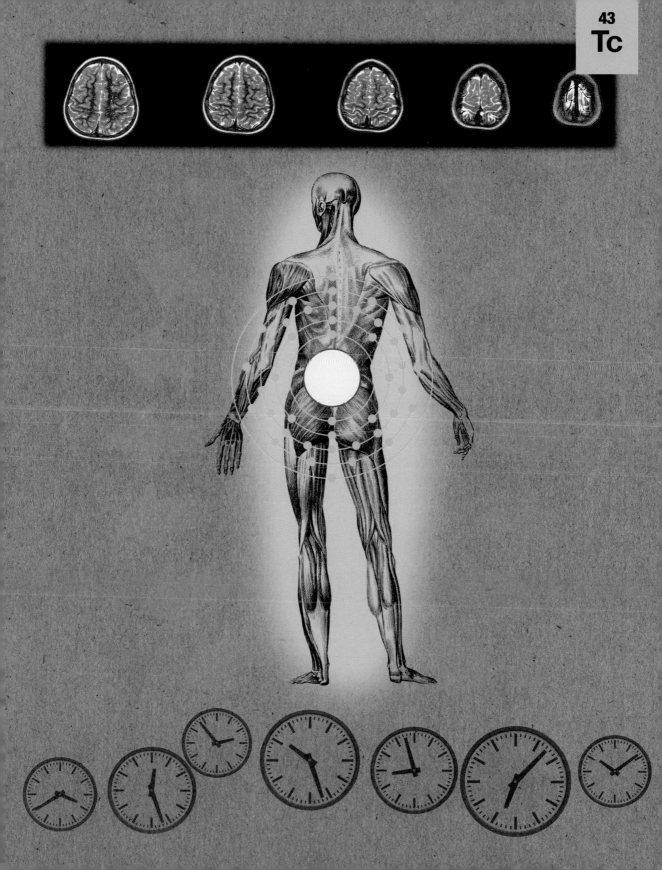

1 February 1905
Born in Tivoli, near Rome

1922
Attends University of
Rome, studying first
engineering then physics

1928
Achieves physics
doctorate supervised
by Enrico Fermi

1928
Serves a year in the
Italian army

1932
Appointed assistant
professor at the
University of Rome

1936
Becomes director of the
physics laboratory at the
University of Palermo

1937
Isolates the element
technetium

1938
Dismissed from his post
under anti-semitic laws
in Italy

1940
Isolates the element
astatine. Discovers
plutonium-239 is
fissionable

1943–46
Appointed a group leader
in the Manhattan Project
at Los Alamos National
Laboratory

1959
Wins Nobel Prize for
Physics

1972
Returns to Rome from
United States as a
professor of nuclear
physics

22 April 1989
Dies of a heart attack

EMILIO SEGRÈ

By the 1920s, a handful of elusive elements remained to be discovered on Dmitri Mendeleev's periodic table. Given the radioactive nature of these elements – and more that Mendeleev did not predict – it was time for physicists to step forward in the search. Nobel Prize winner Emilio Segrè was a pioneering atomic and nuclear physicist who discovered artificially created elements that could not be found on the earth.

Born in 1905, Segrè grew up in Tivoli and studied physics at the University of Rome. He completed his doctorate in 1928 under the supervision of Enrico Fermi, one of the leading nuclear physicists of the 20th century. During the 1930s, Segrè was part of Fermi's young team at the University of Rome that became famous for groundbreaking discoveries in neutron bombardment, in particular the production of slow neutrons that would later be used to trigger nuclear fission reactions.

In 1936, Segrè was appointed director of physics at the University of Palermo, where he used his experience in Rome effectively. Scientists knew that there was a 'missing' element in the periodic table under manganese and set out to predict its properties. However, element 43 proved difficult to find.

In 1937, at the University of California, Berkeley, scientists sent Segrè and mineralogist Carlo Perrier a strip of molybdenum that had been bombarded with deuterons in a cyclotron and was producing anomalous radioactivity. Segrè proved that the radiation emitted was produced by technetium, and he entered the record books by identifying it as the first artificially synthesized chemical element. Technetium has a half-life of 4 million years, so any produced when the planet was formed 4.57 billion years ago would be long gone.

Segrè was Jewish and on a research trip to California in 1938 was dismissed from his post at Palermo by Benito Mussolini's fascist government. Working at Berkeley during wartime, he helped to discover the element astatine and the isotope plutonium-239, which was fissionable. In 1943, Segrè became a group leader in the Manhattan Project, developing the atomic bomb that used plutonium-239 with such deadly consequences on Nagasaki, Japan, on 9 August 1945.

Segrè became a US citizen and taught at Berkeley until 1972. Working with the American physicist Owen Chamberlain, he discovered the antiproton (a subatomic antiparticle) and the duo shared the Nobel Prize for Physics in 1959 for their achievement.

SILVER
the 30-second element

Among metals, silver stands

supreme in three ways: it is the best conductor of electricity, it is the best conductor of heat and it gives the best reflectance (a technical measure of how well a surface reflects). These features are exploited commercially in grinding wheels, electronics and mirrors. Silver solder is used to attach industrial diamonds to grinding wheels, because it dissipates the heat generated more effectively. Silver is widely used for electrical and electronic devices, because it makes and breaks electric circuits cleanly, and, in addition to mirrors, is used for trophies and special tableware. The chief silver ore is acanthite (silver sulphide), but most silver is obtained as a by-product in the refining of copper and lead. Silver salts are sensitive to light and were an essential part of photographic film. Now they feature in reactive sunglasses. Sunlight converts colourless silver ions (Ag^+) to metallic silver by taking an electron from a copper atom, and the glass darkens; when the light fades, the electron returns to the copper. Silver is deadly to bacteria and viruses, and silver nitrate used to be applied to wounds as an antiseptic. It is now added to paints to keep surfaces free of disease pathogens.

3-SECOND STATE
Chemical symbol: Ag
Atomic number: 47
Named: From Anglo Saxon *siolfur*; Ag from Latin *argentum*

3-MINUTE REACTION
Silver is a member of group 11 of the periodic table, the coinage metals. It is stable to oxygen and water, but dissolves in sulphuric and nitric acids. The metal is slowly attacked by sulphur compounds in the air that form black silver sulphide. Silver nitrate was known to the ancients as lunar caustic. This salt was remarkably soluble in water and much used as a caustic; silver chloride – completely insoluble – was used in precipitating silver.

RELATED ELEMENTS
See also
COPPER (Cu 29)
page 76

GOLD (Au 79)
page 88

3-SECOND BIOGRAPHIES
JOSEPH NICÉPHORE NIÉPCE
1765–1833
French inventor who took the first ever photograph in 1816, using silver chloride

JOHN WRIGHT
1808–44
British doctor who discovered how to silver plate other metals, in 1840

CARL FRANZ CREDÉ
1819–92
German gynaecologist who introduced silver nitrate drops in 1884 to kill a virus in babies

30-SECOND TEXT
John Emsley

Silver would take gold medal in a competition for the best conductor and reflector. Unless you're an Olympic athlete, you're perhaps most likely to handle it in silver-plated cutlery.

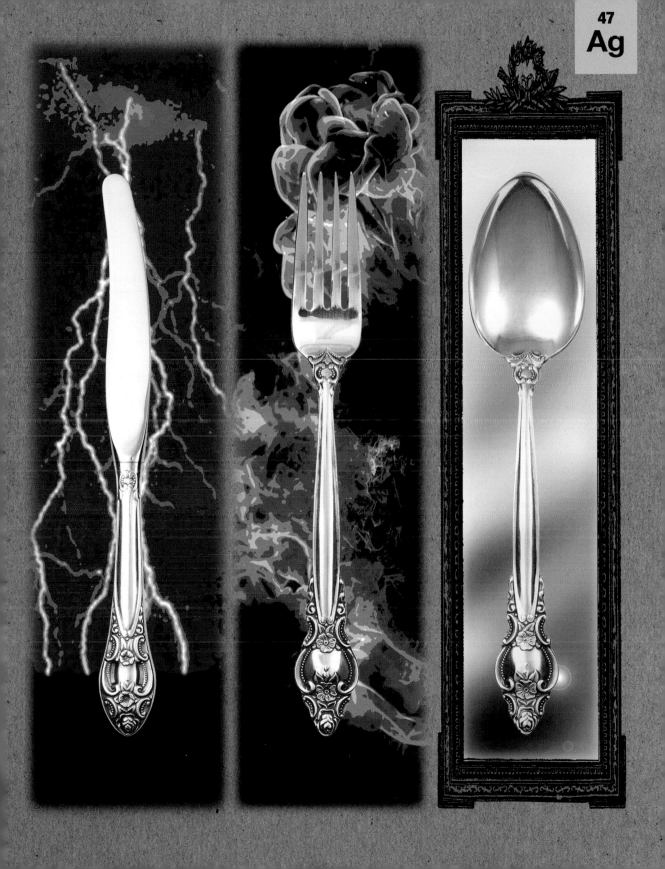

HAFNIUM

the 30-second element

Hafnium is a silvery, ductile

metal with corrosion-resistant properties. A remarkable aspect of element 72 is the number of the priority disputes over its discovery. One of the first scientists who believed he had discovered the element was French inorganic chemist Georges Urbain, in 1911. Then English physicist Henry Moseley established an X-ray method that provided a definitive way of checking the atomic number of any particular element and showed that Urbain had not isolated element 72. A few years later, however, Urbain revived his claim and was supported by the popular press, especially in France and Britain. This was shortly after the First World War, when rivalries were strong between England and France on one side and the Germanic nations on the other; Denmark was not strictly a Germanic nation, and neither of the two scientists, Dutchman Dirk Coster and Hungarian George de Hevesy, who discovered hafnium in Denmark, were Danish. Nevertheless, they were subjected to press criticism and ridicule until they produced X-ray evidence. They were eventually declared discoverers of the new element. It is used to make control rods for nuclear reactors and is present in many high-tech alloys used in the space and computer industries.

Chemical symbol: Hf
Atomic number: 72
Named: From *Hafnia*,
the Latin name for
Copenhagen, the city in
which it was discovered

3-MINUTE REACTION

Hafnium is not a particularly rare element, but it was difficult to extract because it is so similar to the element zirconium that lies directly above it in the periodic table. The two elements typically occur together in minerals such as zircon or $ZrSiO_4$. Hafnium absorbs neutrons well and is used for that purpose in nuclear reactors.

RELATED ELEMENTS
See also
TITANIUM (Ti 22)
page 71

ZIRCONIUM (Zr 40)
page 71

RUTHERFORDIUM (Rf 104)
page 71

3-SECOND BIOGRAPHIES
GEORGES URBAIN
1872–1938
French chemist who falsely claimed in 1911 to have discovered element 72, which he called celtium

GEORGE DE HEVESY
1885–1966
Hungarian radiochemist, co-discoverer of hafnium

DIRK COSTER
1889–1950
Dutch physicist, co-discoverer of hafnium

30-SECOND TEXT
Eric Scerri

Hafnium is extracted from zirconium minerals for use in alloys and the control rods needed within nuclear power stations.

72
Hf

RHENIUM

the 30-second element

The element rhenium lies two places below manganese in group 7 of the periodic table. Its existence, as well as that of an element above it, were predicted by Russian chemist Dmitri Mendeleev in 1869. Rhenium was finally discovered in 1925 by Walter Noddack, Ida Tacke (later Noddack) and Otto Berg in Germany. After an extraction of heroic proportions, they obtained around 1 g ($\frac{1}{25}$ oz) of rhenium by processing about 660 kg (1,450 pounds) of the ore molybdenite. Until quite recently, no mineral containing rhenium combined only with a non-metal had been found. In 1992, however, a team of Russian scientists discovered rhenium disulphide at the mouth of a volcano on an island off the east coast of Russia. In contrast to many other metals, rhenium does not undergo transformation from ductile to brittle as its melting point is approached. It retains a very high strength at high temperatures in addition to very good ductility, making it an ideal choice for high-temperature applications. Recently, a simple compound, rhenium dibromide, has attracted attention as one of the hardest known substances; unlike other superhard materials, it does not have to be manufactured under high pressure.

3-SECOND STATE
Element symbol: Re
Atomic number: 75
Named: From *Rhenus*
(Latin for the river 'Rhine')

3-MINUTE REACTION
Rhenium shows the largest range of oxidation states of any known element, namely -1, 0, +1, +2 and so on all the way to +7, the last of which is its most common oxidation state. It is also the metal that led to the discovery of the first metal-to-metal quadruple bond as found in 1964 in the rhenium ion $[Re_2Cl_8]^{2-}$.

RELATED ELEMENTS
See also
BOHRIUM (Bh 107)
page 71

MANGANESE (Mn 25)
page 71

TECHNETIUM (Tc 43)
page 78

3-SECOND BIOGRAPHIES
WALTER & IDA NODDACK
1893–1960 &
1896–1978
German chemists,
co-discoverers of rhenium

ALBERT COTTON
1930–2007
American chemist, prepared the first metal compound with a quadruple metal-metal bond using rhenium

30-SECOND TEXT
Eric Scerri

The very hard-wearing, silvery metal rhenium resists corrosion and has been used in electrical contacts and in the nibs of fountain pens. It can be made into wire and foil.

GOLD

the 30-second element

Technically a transition metal

(part of a large block in the centre of the periodic table), gold is used above all in jewellery and as currency – reflecting its ease of working, its rarity and its attractive shine. It differs from the usual silvery colour of metals because some of its electrons move so fast (close to the speed of light) that relativistic effects change the shape of their orbits, altering the energy of the photons they absorb and re-emit. Because gold is so dense, practically all the earth's gold is thought to be deep within the planet. The metal we dig up arrived later, when gold-bearing asteroids and meteorites hit the earth's surface. It has been estimated that all the gold ever mined would form a block around the size of a small office block – 8,000 m^3 (282,500 cu ft). From earliest times, gold has found its way into jewellery, and this still accounts for around 50 per cent of production; another 40 per cent is transformed into gold bars and coinage. The remainder has the most practical usage: because it doesn't oxidize in air and is a great conductor, gold is often used for circuit boards, plugs and electrical contacts.

3-SECOND STATE
Chemical symbol: Au
Atomic number: 79
Named: From the old German *ghol* (prefix for yellow)

3-MINUTE REACTION
Gold is not highly reactive, which is why it stays shiny, not oxidizing in air, but it will dissolve in aqua regia, a mix of concentrated nitric and hydrochloric acids. It is classed as a noble metal alongside silver, platinum and others, because filled bands in its electronic structure give it low reactivity. It can react, though, typically producing compounds such as gold chlorides AuCl and Au_2Cl_6.

RELATED ELEMENTS
See also
COPPER (Cu 29)
page 82

SILVER (Ag 47)
page 82

3-SECOND BIOGRAPHY
ARCHIMEDES
C. 287–C. 212 BCE
Greek philosopher who tested gold's density by dunking it into water

PEKKA PYKKÖ
1942–
Finnish quantum chemist who has predicted several new compounds of gold

30-SECOND TEXT
Brian Clegg

Gold has been known for at least 6000 years. Its lure, which finds it coating everything from Olympic medals to Oscars, remains its luxurious, glittery scarcity.

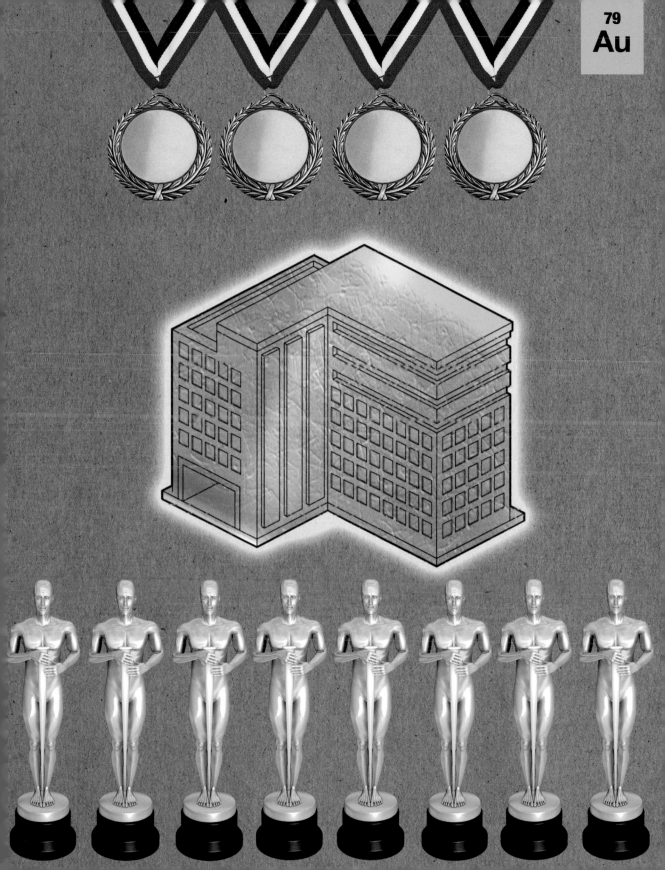

MERCURY

the 30-second elements

Mercury is the only liquid metal
and, with bromine, one of only two elements that is liquid at room temperature. Its liquid nature makes it distractingly beautiful: the Islamic rulers of medieval Spain placed mercury pools in their gardens in which visitors could dabble their fingers. The element is usually obtained from its ore cinnabar or vermilion (mercury sulphide), which is also the pigment used to produce the red colour used in some Hindu rituals. Mercury has been used as a medicine for thousands of years in forms such as the laxative calomel and the disinfectant mercurochrome; more strongly reactive compounds were used to treat syphilis. Mercury is especially favoured in Chinese medicine. The element is nevertheless highly poisonous. Mercury used to treat animal fur in hat-making produced acute psychological as well as physical symptoms of illness, inspiring the phrase 'as mad as a hatter' and the character of the Hatter in Lewis Carroll's 1865 novel *Alice's Adventures in Wonderland*. Less toxic substitutes are being found for many of mercury's uses, in measurement instruments, valves, switches and dental amalgams. However, other applications such as in energy-saving compact fluorescent bulbs are increasing demand for the element.

3-SECOND STATE
Chemical symbol: Hg
Atomic number: 80
Named: From its alchemical and astrological links to the planet Mercury

3-MINUTE REACTION
The chemical behavior of mercury sulphide was of great interest to alchemists, who hoped that combining sulphur with mercury might produce gold. Later, chemists saw that this reversible reaction (heating mercury and sulphur leads to the sulphide; heating again makes it break up into its constituents) provided a clue that elements can be neither created nor destroyed. British natural philosopher Joseph Priestley exploited the similar reactions of mercury oxide in his experiments with oxygen in 1774.

RELATED ELEMENTS
See also
ZINC (Zn 30)
page 51

COPERNICIUM (Cn 112)
page 92

3-SECOND BIOGRAPHIES
EVANGELISTA TORRICELLI
1608–47
Italian inventor of the mercury barometer in 1643

DANIEL FAHRENHEIT
1686–1736
Dutch-German-Polish inventor of the mercury thermometer

ALEXANDER CALDER
1898–1976
American creator of 1937 artwork 'Mercury Fountain'

30-SECOND TEXT
Hugh Aldersey-Williams

Poisonous but beautiful, mercury caused the madness of hatters in the work of Lewis Carroll. Today, it is used safely in instruments and fluorescent bulbs.

COPERNICIUM

the 30-second element

Right now it probably doesn't

exist. Copernicium is one of a group of superheavy elements made artificially in a particle accelerator by colliding ions into a heavy-metal target; and, like the other superheavy elements, it is radioactive and decays very quickly. The longest-lived isotope, copernicium-285, has a half-life of just 29 seconds. These elements are made atom by atom, and, in total, just 75 atoms of copernicium have been detected so far. The element was first produced in 1996 by firing zinc ions into lead at the GSI Center for Heavy Ion Research in Darmstadt, Germany – the birthplace of several other artificial elements. The German claims were not officially recognized until 2009, when the team proposed to name the new element after Polish astronomer Nicolaus Copernicus. That name wasn't accepted until 19 February 2010, the 537th anniversary of Copernicus's birth. By that time copernicium had been synthesized by other groups in Russia and Japan. Investigating copernicium's chemical properties is greatly challenging, given so little material (a few atoms) and so short a time (just a few seconds). Its place in the periodic table suggests it should be similar to mercury, forming chemical bonds to gold – and that's what experiments so far seem to confirm.

3-SECOND STATE
Chemical symbol: Cn
Atomic number: 112
Named: After the Polish astronomer Nicolaus Copernicus (1473–1543)

3-MINUTE REACTION
Copernicium is the heaviest of the elements in group 12 of the periodic table – which also includes zinc, cadmium and mercury. In a copernicium atom, the huge nucleus – with a very large positive charge – distorts energy levels through the electron shells because of the effects of special relativity with some of the electrons moving so fast that they gain mass. This has knock-on effects that may make copernicium exhibit the behaviour of a noble gas.

RELATED ELEMENTS
See also
CADMIUM (Cd 48)
page 71

MERCURY (Hg 80)
page 90

ROENTGIUM (Rg 111)
page 71

3-SECOND BIOGRAPHY
SIGURD HOFMANN
1944–
German chemist, leader of the team that discovered copernicium in 1996

30-SECOND TEXT
Philip Ball

Superheavy element copernicium is the product of experiments in an ion accelerator. The first time it was produced, just one single atom of copernicium-277 was created.

METALLOIDS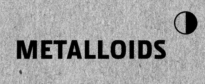

borax Sodium borate, $Na_2B_4O_7 \cdot 10H_2O$, a salt of boric acid used in cosmetics and detergents, in glazes in the ceramics industry, and since ancient Egyptian times as a flux in metalworking.

buckyball Also known as buckminster-fullerene, a spherical carbon molecule with formula C_{60}. It has a structure like a football, containing 12 pentagons and 20 hexagons with a carbon atom at each vertex (corner/intersection). It is an example of a fullerene, a type of carbon molecule in the shape of a sphere or tube. Buckminsterfullerene was the first fullerene discovered and was produced in 1985 by Richard Smalley, Robert Curl, James Heath, Sean O'Brien and Harry Kroto. The name of the molecule is a tribute to American polymath R. Buckminster Fuller (1895–1983) because its shape recalls the geodesic dome that he developed.

dopant Trace amount of impure material added to a pure material; used, for example, in controlling the conductivity of a semiconductor.

doping In general, the addition of impurities. In the context of semiconductors, doping means adding very small amounts of impure materials to pure semiconductors to control their conductivity. For example, adding 10 atoms of boron per million atoms of silicon increases the conductivity of silicon by a factor of 1000. Doping agents are also added to phosphors to produce a specific colour in the glow they give off when stimulated by ultraviolet light or electrons. The lanthanide element europium is used in this way.

electron hole A theoretical concept used by chemists, physicists and electronic engineers to describe a gap that an electron could occupy in an atom. The gap will attract an available electron.

electroplating Coating a material with a very thin layer of metal using electrolysis such as in chrome or gold plating. The material being coated must be able to conduct electricity; it is either a conductive metal or a material made conductive by techniques including covering with lacquer or graphite.

flux Cleaning or purifying substance used in metalworking, for example, to remove oxides from the surfaces of molten metals. Fluxes are also used in smelting ores to make metals flow more easily.

hydride Chemical compound in which hydrogen and another element are combined.

N-type material A material, typically a semiconductor, treated with impurities (dopants) to make it have more conductive electrons than electron holes. An example is silicon doped with arsenic or phosphorus.

P-type material A material, typically a semiconductor, treated with impurities (dopants) to make it have more electron holes than conductive electrons. An example is silicon doped with boron or aluminium.

semiconductor Material with electrical conductivity between that of a conductor (which allows an electrical charge to flow easily through it) and an insulator (which prevents an electrical charge flowing easily through it). Semiconductors have good conductivity in certain conditions, depending on factors such as temperature, magnetic fields or the addition of impurities (doping). Some semiconductors are elements, for example, tin, silicon and germanium; other semiconductors are compounds, for example, gallium arsenide.

semi-metal Another term for metalloid, an element that is intermediate between a metal and a non-metal. Examples include aluminium and germanium.

sublimation Passing directly from solid to gaseous state without becoming a liquid.

METALLOIDS
Metalloids are found in groups 13–16 of the periodic table. They have characteristics of non-metals and of metals. Metalloids such as boron, germanium and silicon are semiconductors and are used in computer chips.

Metalloids

	Symbol	Atomic Number
Boron	B	5
Silicon	Si	14
Germanium	Ge	32
Arsenic	As	33
Antimony	Sb	51
Tellurium	Te	52
Polonium	Po	84

BORON

the 30-second element

Boron is the third lightest solid non-metal, very different from aluminium and the other metals in its group. It was first obtained, as light but very hard, dark grains, by British chemist Humphry Davy in London, and – independently of Davy – by the French chemists Louis-Joseph Gay-Lussac and Louis-Jacques Thénard, working together, in Paris in 1808. Boron is scarce but widely distributed in the earth's crust, and there are a few rich deposits of borates, in which the element is combined with oxygen and calcium or sodium; the latter (borax) has been used since ancient times as a flux for working with molten metals. Boron is essential to plant life and as a micronutrient for animals, toxic only in excess. It combines with carbon in a very hard ceramic, used in tank armour, bullet-proof vests and many industrial applications, and also to shield nuclear reactors. One of boron's compounds with nitrogen has the structure of diamond and is almost as hard, but is more heat resistant and therefore a valuable abrasive. There is also a form like graphite. Boric acid is an antiseptic and is one of the few things that can kill cockroaches.

SILICON

the 30-second element

Every valley is really a silicon

valley. Silicate minerals, containing frameworks of silicon and oxygen atoms, comprise most of the earth's crust; silicon is the crust's second most abundant element, after oxygen. However, despite this ubiquity, pure silicon was not isolated until 1824, because it is hard to separate it from oxygen. Much of mineral chemistry centres on the various ways that silicate ions – tetrahedra with silicon at the centre and oxygen atoms at the four corners – can be joined into orderly, crystalline networks that house metallic elements such as sodium and calcium. In glass, this silicon-oxygen network is melted and refrozen into static disorder. Today, the ultra-pure silicon needed for microelectronics is mostly made by electrolysis of low-purity silicon or its compounds, followed by melting and controlled crystallization to make the near-perfect crystals needed for silicon wafers. Silicon is a semiconductor. It contains just a few 'free' electrons that can carry an electrical current. The number of these mobile electrons, and thus the electrical conductivity, can be finely tuned by adding small quantities of impurities ('dopants') such as boron and phosphorus; this is why silicon is so useful for making microelectronic devices such as transistors.

3-SECOND STATE
Chemical symbol: Si
Atomic number: 14
Named: From Latin *silex* ('hard rock')

3-MINUTE REACTION
Silicon is regularly confused with silicone, a class of polymers whose backbones are chains of alternating silicon and oxygen atoms. In silicones, each silicon atom has hydrocarbon appendages. Developed in the early 20th century, silicones range, like hydrocarbon polymers, from oils to tough plastics, finding uses such as lubricants, sealants, adhesives, electrical insulation, cookware, and – controversially – breast implants. The ability of silicon to form polymers analogous to hydrocarbons sparks speculation about silicon-based alien life.

RELATED ELEMENTS
See also
SELENIUM (Se 34)
page 135

OXYGEN (O 8)
page 142

3-SECOND BIOGRAPHIES
JÖNS JAKOB BERZELIUS
1779–1848
Swedish chemist, the first person to isolate fairly pure silicon, in 1824

VICTOR MORITZ GOLDSCHMIDT
1888–1947
Swiss mineralogist who deduced the crystal structures of many silicate minerals

30-SECOND TEXT
Philip Ball

As a key ingredient of rocks and stones, silicon has a history of human use reaching back to the Stone Age, despite modern associations with computing and hi-tech industries.

GERMANIUM

the 30-second element

When Russian chemist Dmitri
Mendeleev drew up the periodic table of
elements, he found gaps where he predicted
unknown elements would sit. One, which he
called 'eka-silicon', proved to be germanium.
Mendeleev predicted its atomic weight, its
density and even that this semi-conducting
metalloid would be grey. German chemist
Clemens Winkler, who in 1886 discovered
the actual element, intended to call his find
'neptunium', but discovered that another
element had been given that name, so went
for the Latin term for the newly formed
Deutschland. Ironically, neptunium came free
again, because the other discovery proved to
be an error. Germanium found its niche in the
20th century as the leading component of early
solid-state electronics, replacing expensive and
fragile glass valves (vacuum tubes). Until the
1970s, germanium transistors and diodes were
common, but silicon took over the mass market
– partially because silicon is so cheap (the raw
material is sand), and also because it is more
effective as a semiconductor; germanium had
got its head start because sufficiently pure
silicon was not widely available before.
However, this hasn't meant the disappearance
of germanium from electronics. Fibre-optic cable
equipment and night vision goggles still make
use of this robust semiconductor.

3-SECOND STATE
Chemical symbol: Ge
Atomic number: 32
Named: From Latin
Germania (name for
the German region)

3-MINUTE REACTION
Germanium first found
large-scale use as a
replacement for the
diode valve. The valve acts
as a one-way mechanism,
only allowing current to
flow in one direction;
a semi-conductor like
germanium provides a
solid-state equivalent.
When impurities are
added, germanium can
become an electron
donor (n-type material)
or electron acceptor
(p-type); by combining
strips of the different
types of germanium,
electrons can be limited
to travelling in a single
direction.

RELATED ELEMENTS
See also
SILICON (Si 14)
page 100

ARSENIC (As 33)
page 104

3-SECOND BIOGRAPHIES
DMITRI MENDELEEV
1834–1907
Russian chemist who predicted
the existence of germanium
from his periodic table

CLEMENS WINKLER
1838–1904
German chemist who
discovered germanium in 1886

30-SECOND TEXT
Brian Clegg

*While rhenium is
named for the River
Rhine, germanium –
initially called 'eka-
silicon' by Dmitri
Mendeleev (right) –
takes its title from the
country of Germany.*

ARSENIC
the 30-second element

Arsenic is a semimetal, best known as its oxide (white arsenic), used as a poison over many centuries. White arsenic was scraped from the flues of copper refineries when ores rich in arsenic were smelted and, despite its toxicity, became popular in medicine from 1780 onward – as 'Dr. Fowler's Solution', which was prescribed for all kinds of ailments but with little benefit. The arsenic medicine Salvarsan, discovered in 1909, cured parasitic infections of the blood such as syphilis, and white arsenic reappeared as the medicine arsenic trioxide, now used to treat leukemia. Accidental arsenic poisoning appeared to be a threat in the 19th century due to the green pigment copper arsenite, used in wallpapers; when it became damp, it could give off trimethylarsine vapour and was thought to cause arsenic poisoning, maybe even of the deposed French emperor Napoleon in 1821. Then, in 2005, it was shown that this gas is not particularly toxic. Arsenic-based weedkillers and wood preservatives have now been phased out, and today arsenic is more likely to be used as the semiconductor gallium arsenide (GaAs). Arsenic is present in foods, such as prawns, but in a form that poses no threat to health.

3-SECOND STATE
Chemical symbol: As
Atomic number: 33
Named: After Greek *arsenikon* (the mineral yellow orpiment)

3-MINUTE REACTION
A semimetal in group 15 of the periodic table, arsenic can bond to three other atoms, as in the deadly gas arsine (AsH_3); or to five atoms, in compounds such as the pentachloride ($AsCl_5$). It has two oxides, As_2O_3 and As_2O_5. Arsenic forms two kinds of salts: arsenites (which have the negative ion AsO_3^{3-}) and arsenates (negative ion AsO_4^{3-}). When it is strongly heated, arsenic sublimes (which passes from solid to gas without becoming liquid) at 616°C (1,141°F).

RELATED ELEMENTS
See also
PHOSPHORUS (P 15)
page 144

GALLIUM (Ga 31)
page 120

ANTIMONY (Sb 51)
page 106

3-SECOND BIOGRAPHIES
ALBERTUS MAGNUS
1193–1280
German friar who was the first person to separate arsenic

CARL WILHELM SCHEELE
1742–86
German-Swedish chemist who discovered the copper arsenite dye called Scheele's green in 1775

PAUL EHRLICH
1854–1915
German doctor who discovered Salvarsan as a cure for syphilis

30-SECOND TEXT
John Emsley

Famous for its poisonous oxide, it has also been used for more than two centuries in medicine and is today used in treating leukaemia.

ANTIMONY

the 30-second element

Assyrians and ancient Egyptians decorated their eyes with the black mineral stibnite (antimony sulphide), from which antimony gets its chemical symbol Sb. To the Assyrians, stibnite was *guhlu*, which became Arabic *kohl* – a word still in use for eyeliner; *al-kohl* came to be used for any fine powder, and then for distilled liquids, transferring finally to the English word 'alcohol'. Stibnite could cure eye infections and was considered a powerful medicine. Medieval alchemists revered its medicinal powers, and the pseudonymous chemist Basil Valentine celebrated it in a 1604 book *The Triumphal Chariot of Antimony*. The 17th century saw the Antimony Wars, in which French chemists argued over whether antimony was poison or cure. In fact, the element is quite toxic, and Basil Valentine claimed that when he administered it in a monastery, some of the monks died – accounting for the probably apocryphal etymology anti-monachos: 'anti-monk'. Antimony was a favourite slow poison among Victorian murderers. Pure antimony, first made in the 16th century, looks like a silvery metal, but it is a metalloid, with properties in between a metal and non-metal. It conducts electricity but is soft, and it is mostly used now in alloys of lead and tin for solders and lead-acid battery electrodes.

3-SECOND STATE
Chemical symbol: Sb
Atomic number: 51
Named: Disputed etymology – possibly Arabic or Greek

3-MINUTE REACTION
Antimony's toxicity can cause vomiting when it is ingested. This led to its use as a medieval 'purgative' – it was considered a good way to expel disease from the body. By the same token it has laxative properties, and constipation caused by the poor diet of the Middle Ages was cured by swallowing tablets of pure antimony. The tablets were not absorbed by the body, but were excreted – and recovered for reuse.

RELATED ELEMENTS
See also
ARSENIC (As 33)
page 104

BISMUTH (Bi 83)
page 117

3-SECOND BIOGRAPHIES
JOHN OF RUPESCISSA
c. 1310–c. 1362
French alchemist who considered antimony central to alchemical medicine

JOHANN THÖLDE
1565–1614
German publisher, and probable author under the name Basil Valentine, of *The Triumphal Chariot of Antimony* (1604)

30-SECOND TEXT
Philip Ball

One theory about the death at only 35 of Austrian composer Wolfgang Amadeus Mozart (top right) is that he was poisoned by excessive 'medicinal' doses of antimony.

TELLURIUM

the 30-second element

Tellurium was the *metallum problematum* ('problem metal') for 16 years after its discovery in 1782 in gold ores from Transylvania (now part of Romania). The properties of the ores indicated that the substance had metallic and non-metallic properties, so it was also called *aurum paradoxum* ('paradoxical gold'). Tellurium was identified as an element in 1798. In 1834, Swedish chemist Jöns Jacob Berzelius decided it was a metal, but one that belonged in the same group as the non-metals sulphur and selenium due to the similarities of their compounds; it is now recognized as a metalloid. Tellurium is about 15 per cent less dense than iron and marginally harder than sulphur. It is easily pulverized and too brittle for structural uses; its primary use is as an additive to improve the machining qualities, or workability, of various metals, chiefly stainless steels and copper. Occasionally found native as elemental crystals, it combines easily with most other elements to form tellurides. The semiconducting materials bismuth telluride and lead telluride are used in thermoelectric devices as sources of electricity or for cooling. Cadmium telluride thin-film solar cells, used to generate electric power, are amongst the lowest-cost type of solar cell.

3-SECOND STATE
Chemical symbol: Te
Atomic number: 52
Named: From Latin *tellus* ('earth'/'the earth')

3-MINUTE REACTION
Tellurium is extremely rare in the earth's crust, partially due to the fact that its formation of a volatile hydride caused the element to be lost to space as a gas during the hot nebular formation of the earth. It is highly toxic. In the body, it is partially metabolized to the gas dimethyl telluride, which can be smelled in 'tellurium breath', an unpleasant garlic-like odour identifiable in the breath of people who have been exposed to tellurium.

RELATED ELEMENTS
See also
COPPER (Cu 29)
page 76

ARSENIC (As 33)
page 104

BISMUTH (Bi 83)
page 117

3-SECOND BIOGRAPHIES
FRANZ-JOSEPH MÜLLER VON REICHENSTEIN
1740–1826
Austrian mining engineer who discovered tellurium in gold ore in 1782

MARTIN HEINRICH KLAPROTH
1743–1817
German chemist who named tellurium in 1798

30-SECOND TEXT
Jeffrey Owen Moran

'Garlic breath' can have consequences more serious than social embarrassment, it may mean you have tellurium poisoning. Tellurium can damage your liver and nerves.

7 November 1867
Born in Warsaw

1891
Moves to Paris to study mathematical sciences and physics at the Sorbonne

1895
Marries Pierre Curie

1897
Daughter Irène born

1898
The Curies discover polonium and radium

1903
Becomes the first woman to receive a doctorate of science at the Sorbonne

1903
Jointly awarded the Nobel Prize for Physics with Pierre and Becquerel

1904
Daughter Ève born

1906
Pierre dies in a road accident

1906
Becomes the first ever female professor at the Sorbonne

1911
Wins a Nobel Prize for chemistry

1914
Appointed a director of the newly founded Radium Institute in Paris

1934
Dies in France of leukemia

1995
The Curies' remains are reinterred in the Panthéon, Paris

MARIE CURIE

Millions of people across the world owe their lives to the pioneering work of Marie Curie. She overcame poverty, sexism and illness to pursue her passion for physics and, along with her husband Pierre, discovered radium, which would be used in radiation therapy to treat cancer. Marie was the first woman to win a Nobel Prize and dedicated her life to study of the therapeutic possibilities of radiotherapy.

Born Maria Sklodowska in Warsaw in 1867, Marie showed an early gift for physics. Women were banned from attending university in Warsaw at that time, so she had to work as a governess to scrape together the funds to study at the Sorbonne in Paris. In 1894, she met Pierre Curie, a professor of physics, and their marriage marked the beginning of a groundbreaking scientific collaboration.

The scientific community was hugely excited in 1896 when Henri Becquerel discovered that uranium emitted radiation. The Curies were intrigued and began to examine uranium ore (pitchblende). They deduced that the ore was more radioactive than uranium itself, and that it must, therefore contain other radioactive elements. After years of painstaking research – suffering ill health and exhaustion – the Curies made history by discovering the highly radioactive elements polonium and radium. No one would dare handle radium now, but the Curies were determined to experiment and found that when applied to human flesh radium could burn and wound. It was this discovery that led to the use of radium in the treatment of cancerous tumours.

Awards and honours began to flood in, and in 1903 Marie made history by becoming the first woman to win a Nobel Prize, which she shared with Becquerel and Pierre. Pierre died in a road accident in 1906, but Marie continued their work at the Sorbonne. She made history again in 1911 by being awarded a second Nobel Prize, this time for chemistry, in recognition of her achievements in the study of radium.

During the First World War, she pioneered the use of mobile radiography units to diagnose injuries, and drove the ambulances herself. After the war, her health began to decline and she died of leukaemia in 1934; this may have been caused by exposure to radioactive material. Marie's name will always be synonymous with the fight against cancer and as history's greatest female scientist she has inspired many women to explore the previously male-dominated fields of physics and chemistry.

POLONIUM

the 30-second element

A chemical curiosity from the bottom of the periodic table, polonium is a silvery-white metal of which few people have seen even a speck. This mysterious element is known less for its chemistry than for its physical properties. It was discovered in Paris in 1898 by the Polish-born physicist Marie Curie and her French physicist husband, Pierre, who together isolated it as one of the minor but more radioactive components of pitchblende, a uranium-containing mineral. So radioactive is the metal that it has few uses. However, trace amounts of polonium are incorporated into anti-static devices for the textile, electronics, printing and munitions industries, where sparks would pose a fire or explosion risk. In a remarkable example of fighting fire with fire, the ions produced by polonium's intense radioactivity neutralize any localized build-up of static charge. Polonium became notorious in 2006 when Alexander Litvinenko, a Russian émigré living in London, became mysteriously ill. Doctors at London hospitals took several days to realize that he was suffering from radiation poisoning, the result of swallowing a drink laced with polonium widely believed to have been administered by Russian agents. Litvinenko died in agony three weeks later.

3-SECOND STATE
Chemical symbol: Po
Atomic number: 84
Named: For Poland, birth country of discoverer, Marie Curie

3-MINUTE REACTION
Because natural polonium is so radioactive that an ingot would be hot to the touch, very little is known about its properties – it is just too dangerous to study in ordinary labs. The metal can be dissolved in dilute acids to form salts, electroplated and even distilled under vacuum. The handful of its compounds that have been made – for example, oxide, chloride, bromide – suggest similarities with tellurium and bismuth.

RELATED ELEMENTS
See also
TELLURIUM (Te 52)
page 108

RADIUM (Ra 88)
page 28

URANIUM (U 92)
page 42

3-SECOND BIOGRAPHIES
HENRI BECQUEREL
1852–1908
French physicist who discovered radioactivity in pitchblende

MARIE CURIE
1867–1934
Polish-born physicist who discovered polonium in samples of pitchblende and purified the element

30-SECOND TEXT
Andrea Sella

Polonium was the first element discovered by radiochemical analysis. Its longest-lived isotope has a half-life of 103 years; its most common isotope has a half-life of 138 days.

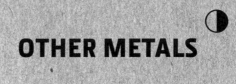

OTHER METALS

alumina Aluminium oxide, Al_2O_3, used in the manufacture of aluminium. Alumina is an example of an amphoteric oxide.

amphoteric Describes an element or compound that can react either as an acid or as a base. The word is derived from the Greek *amphoteroi* (meaning 'both'). Many metalloids and metals – including aluminium, lead and tin – form amphoteric oxides.

bauxite Main ore of aluminium, rock consisting of a combination of hydrous aluminium oxides. Bauxite was named by French geologist Pierre Berthier (1782–1861) after the southern French village of Les Baux-de-Provence, where he first discovered the rock in 1821.

biocide Chemical or microorganism with the capacity to kill living organisms. Herbicides, fungicides, rodenticides and pesticides are examples. Thallium sulphate was widely used in the United States as a rodenticide until the 1960s.

flame spectroscopy Method of determining the amount of an element in a sample. The sample is treated with a flame, and the wavelengths of light emitted by the constituent elements are studied.

fungicide Chemical or organism used to kill or control fungi/fungal spores. The organotin compound tributyltin oxide, $C_{24}H_{54}OSn_2$, is an effective fungicide.

infrared light Part of the electromagnetic spectrum with wavelengths longer than visible light. It is invisible to the unaided human eye. Scientists working with yttrium indium oxide and yttrium manganese oxide have developed a deep blue compound that does not absorb infrared light and so does not get hot in strong sunlight.

iron pyrite A sulphide mineral with the formula FeS_2. It has a metallic lustre and is a pale yellow colour, making it look a little like gold – for this reason it is nicknamed 'fool's gold'. It is used in industry to produce sulphur dioxide and sulphuric acid; thallium is a by-product of the latter process.

lead acetate $Pb(CH_3COO)_2$, a soluble lead salt with a sweet taste, used in solution by the ancient Romans to sweeten wine – they didn't know about its toxicity. Its use as a sweetener continued for many centuries and German composer Ludwig van Beethoven may have died in 1827 from lead poisoning after drinking wine or medical treatments sweetened with lead acetate.

non-ferrous metal Metal that does not contain iron. Aluminium is a prime example.

organolead compounds Chemical compounds in which lead bonds to carbon.

organotin compounds Organic compounds containing tin. They have been used as fire retardants, pesticides and chemical stabilizers over many decades, but their use is now restricted because they have been shown to be highly toxic.

poor metals Another name for the metallic elements in this chapter. They are sometimes also called the post-transition metals.

reflectance Measure of the amount of radiation reflected by a surface, expressed as a ratio of the reflected radiation to the total radiation falling on the surface.

solder Alloy used to join metal surfaces. The solder has a lower melting point than the metals being worked.

tetraethyl lead An organolead compound, $(CH_3CH_2)_4Pb$, added to petrol from the 1920s onwards to improve fuel performance and economy. Its use was discontinued in the 1970s because of concerns about lead poisoning in humans.

OTHER METALS

The elements classified as 'other metals' are in groups 13–15 of the periodic table. They are sometimes called the 'poor metals', have no lustre and are opaque. Compared to the transition metals (Chapter 4), these 'other metals' are softer, have lower melting/boiling points, and have higher electronegativity; their valence electrons (with which they combine with other elements) are found in only the outer electron shell. The elements in the 'other metals' group have much higher boiling points than the metalloids (Chapter 5).

Other Metals

	Symbol	Atomic Number
Aluminium	Al	13
Gallium	Ga	31
Indium	In	49
Tin	Sn	50
Thallium	Tl	81
Lead	Pb	82
Bismuth	Bi	83

ALUMINIUM

the 30-second element

Light, cheap and ubiquitous, aluminium is the first metal to have attained widespread use since the discovery of iron. Aluminium was isolated in the 1820s and introduced as the newest metal at the exposition universelle in Paris in 1855, and it was initially more expensive than gold. It remained an exotic novelty for 31 years until, with the new availability of cheap electric power, American inventor Charles Martin Hall and French scientist Paul-Louis-Toussaint Héroult independently and almost simultaneously discovered the modern method of commercial production from alumina (aluminium oxide) in 1886. Aluminium forms hard, light, corrosion-resistant alloys. It is used extensively in aircraft construction, building materials, consumer durables, electrical conductors and chemical and food-processing equipment. As one of the few metals that retain full silvery reflectance in finely powdered form, it is an important component of silver-coloured paints. Cheap aluminium depends on cheap electricity: it takes three times more electrical energy to produce one ton of aluminium than is needed to make a ton of steel. This is a great case for recycling, which consumes only 5 per cent of the energy needed to produce the metal from its chief ore, bauxite.

3-SECOND STATE
Chemical symbol: Al
Atomic number: 13
Named: From alum (potassium aluminium sulphate), used by dyers to fix colours since antiquity

3-MINUTE REACTION
Stable aluminium, Al-27, is created when hydrogen fuses with magnesium in large stars or supernovae. The most abundant metal, aluminium makes up about 8 per cent by weight of the earth's solid surface. It is about one third the density of iron or copper, and can be drawn into wire or rolled into thin foil. Highly corrosion-resistant and non-magnetic, it possesses excellent conductive properties, although it is inferior to copper in this respect.

RELATED ELEMENTS
See also:
HYDROGEN (H 1)
page 136

IRON (Fe 26)
page 74

COPPER (Cu 29)
page 76

3-SECOND BIOGRAPHIES
HANS CHRISTIAN ØRSTED
1777–1851
Danish chemist who first separated elemental aluminium from its oxide

FRIEDRICH WÖHLER
1800–82
German chemist who first isolated pure aluminium

30-SECOND TEXT
Jeffrey Owen Moran

Aluminium is the most widely used non-ferrous metal, used in aircraft manufacture, building and other industries; its lack of reactivity with food products makes it particularly useful for canning.

GALLIUM

the 30-second element

Gallium is a metal obtained as
a by-product of zinc and copper refining.
Originally predicted by Russian chemist Dmitri
Mendeleev in 1871, it was not until 1875 that Paul
Émile Lecoq de Boisbaudran actually discovered
it. For many years, the only use of gallium was
in low-melting alloys such as gallistan (gallium/
indium/tin), which melts at −19°C (−2°F) and is
used in place of mercury, but today the element
finds its way into many devices essential to
modern living such as mobile phones; although
annual world production is less than 200 tons,
gallium is vital to the world economy. Gallium
sulphide (GaAs) and gallium nitride (GaN) are
excellent semiconductors and are used in
computers, lasers, solar cells and light-emitting
diodes (LEDs). When activated, GaAs – which
has been used for more than 40 years – glows
red; GaN – relatively new – glows blue. It
completed the spectrum of colours provided by
other LEDs, making it possible to generate light
that was perceived as white. These LEDs use
much less electricity than any other form of
lighting. GaN has other benefits: it is stable
at much higher temperatures than other
semiconductors; unlike silicon, it does not rely
on its crystals being perfect. Gallium salts of
the radioactive isotopes gallium-67 and
gallium-68 are used in medical diagnosis.

*Gallium is employed
in computers, mobile
phones and LEDs.
Blue-glowing gallium
nitride is the 'blue'
of blu-ray technology.*

18 April 1838
Born in Cognac, France, to a winemaking family

Childhood
Receives no formal education but is self-taught from textbooks. Establishes his own self-built laboratory at home.

1859
German physicist Gustav Kirchhoff and his compatriot, chemist Robert Bunsen, pioneer spectroscopy – a field that was key to de Boisbaudran

1875
Discovers the element gallium in zinc ore

1876
Awarded the Cross of the Legion of Honour – the highest accolade in France

1879
Awarded Britain's Royal Society Davy Medal for his services to chemistry

1880
Discovers the lanthanide element samarium

1886
Discovers the lanthanide element dysprosium

1894
After argon is discovered, he predicts the existence of the group of noble gases

1895
Ill health begins to limit his studies

1912
Dies in Paris of ankylosis (immobility or fusion) of the joints

PAUL-EMILE LECOQ DE BOISBAUDRAN

Born in 1838, French chemist

Paul-Émile Lecoq de Boisbaudran could have spent his life amongst the sunny vineyards of Cognac, sampling the produce of his family's winemaking business. Instead, he chose to experiment with spirits of a different kind after becoming fascinated with chemistry as a child and setting to work in his home-built laboratory.

De Boisbaudran became intrigued by the field of spectroscopy. This involved heating a solid or gaseous element that would then produce spectral lines – bright and dark bands visible when the light given off was split into a spectrum using a prism. Since each chemical element has its own unique set of spectral lines, the lines serve as a fingerprint by which elements can be identified. De Boisbaudran dedicated ten years to this field of analysis.

Years slaving over a hot Bunsen burner paid off in 1875 when he made his most important discovery. Examining zinc ore taken from the Pyrenees, the chemist was excited to see spectral lines that he had never noticed before.

Russian chemist Dmitri Mendeleev had earlier highlighted an unfilled place in the periodic table beneath aluminium. He called it 'eka-aluminium' and predicted some of its properties. The properties of the new substance corresponded with Mendeleev's predictions – de Boisbaudran had solved the mystery of the missing element. He presented his findings to the French Academy of Sciences in December 1875 and named it gallium (after his homeland's Latin name, Gallia, and according to some accounts as a joke on his own name Lecoq, since *gallus* means 'cock').

Mendeleev was delighted when he heard that gallium's properties were so close to his predictions. However, de Boisbaudran's value for gallium's density (4.9 g/cm^3) differed from Mendeleev's prediction of 6g/cm^3. Far away in St Petersburg, Russia, Mendeleev asked the Frenchman to remeasure the density and he found that Mendeleev was right.

More success followed for de Boisbaudran when he discovered the lanthanide elements, samarium in 1880 and dysprosium in 1886. Gallium remains his most important find; its compounds are found in semiconductors used in computer chips and in the blue lasers at the heart of blu-ray high-definition movie players. So, why not raise a glass of Cognac to the chemist whose find became a key component in the computer I'm writing on and the Kindle that you may be reading?

INDIUM

the 30-second element

As one of the so-called 'poor metals' that appear after the transition metals, indium doesn't live up to the usual expectations for metals: it has a low melting point (157°C/314°F), is soft and doesn't conduct electricity particularly well. However, those can be useful properties in themselves; indium was produced after the Second World War for use in solders and other alloys that melt easily. Indium minerals are rare, but indium appears as an impurity in zinc ores, which are the main source of the element today. However, the element itself is not especially rare – the earth's crust contains about as much indium as mercury. Most indium today is produced (primarily in China) for making indium tin oxide (ITO), a hard, semiconducting material that is transparent in thin layers. This combination of transparency and electrical conductivity is rare and invaluable: ITO is used as a see-through electrode for liquid crystal displays, light-emitting diodes and solar cells. Indium is also distinguished in belonging to a small group of metals, including tin and zinc, that emit a 'scream' (actually more of a crackle) when bent – a noise produced as atoms in the crystalline metal rearrange themselves into mirror-image planes.

3-SECOND STATE
Chemical symbol: In
Atomic number: 49
Named: From 'indigo',
the colour of its distinctive
spectral signature

3-MINUTE REACTION
Indium lives up to its name, a compound of indium, yttrium, manganese and oxygen that is deep blue – similar to ultramarine, but less sensitive to heat and light. This material is now being explored as a blue pigment for demanding uses on roofs and automobiles. Here, there's the advantage that the compound doesn't absorb infrared light, so doesn't get hot in strong sunlight. The main obstacle to this new blue is the cost of indium itself.

RELATED ELEMENTS
See also
GALLIUM (Ga 31)
page 120

THALLIUM (Tl 81)
page 128

3-SECOND BIOGRAPHIES
HIERONYMUS THEODOR
RICHTER & FERDINAND REICH
1824–98 & 1799–1882
German chemists, co-discoverers of indium in 1863; Richter was first to isolate it

REUBEN D. RIEKE
unknown
American chemist who pioneered the use of reactive indium powder in organic chemistry

30-SECOND TEXT
Philip Ball

Like tin, indium emits a high-pitched crackle on being bent. Indium is used in the manufacture of transistors and semiconductors, as well as in LEDs and solar cells.

TIN

the 30-second element

Because tin does not oxidize well in air, it keeps its silvery appearance longer than most metals. This, combined with its ease of working and the ready availability of tin dioxide ore, has meant that it has been used for more than 5,000 years. Tin is rarely employed alone, because at around 13°C (55°F) it converts from a malleable metal to the α-tin allotrope, also known as 'grey tin', a cubic crystalline structure that is brittle and powdery. However, tin has been widely used in alloys (metals that are mixed when molten) since around 3000 BCE, notably as an additive to copper, forming bronze, a durable alloy well suited to weaponry that transformed early civilization. When the main constituent of the alloy is tin with small amounts of copper (and often antimony) added, the result is pewter, a more easily worked metal that has long been used for plates and drinking vessels. Now, tin is most likely to be found in solder and as plating on food tins to prevent corrosion of the steel beneath. Tin also turns up in organotin compounds, bonded to hydrocarbons, notably tributyltin oxide. This chunky molecule with the formula $C_{24}H_{54}OSn_2$ is widely used as a wood preservative, thanks to its effectiveness as a biocide and fungicide.

3-SECOND STATE
Chemical symbol: Sn
Atomic number: 50
Named: From Old English *tin* and Germanic *zinn*, names for the metal

3-MINUTE REACTION
A tin atom has 50 positively charged protons in its nucleus. This is a 'magic number' (according to the theory of magic nuclear numbers) that indicates particular stability. The nucleus is more strongly bound than usual, making it less likely for nuclear decay to occur. Because of this, tin has many stable isotopes – at ten, the most of any element in the periodic table, running from tin-112 all the way to tin-124.

RELATED ELEMENTS
See also
SILICON (Si 14)
page 100

LEAD (Pb 82)
page 130

3-SECOND BIOGRAPHY
PETER DURAND
18th to 19th century
Englishman who patented the tin can in 1810

30-SECOND TEXT
Brian Clegg

Tin cans – often called 'tins' in British English – are often made of tinplate (tin-coated steel). Familiar from the alloys bronze and pewter, tin is also used in solder.

THALLIUM

the 30-second element

Thallium can be murder, as it was
in Agatha Christie's novel *The Pale Horse*. Toxic
to humans, thallium accumulates slowly in
tissues, around 1 g ($\frac{1}{25}$ oz) being enough to
cause death in about two weeks. Odorless and
tasteless, it gives no warning of its presence,
then causes rapid hair loss and severe nervous
and gastrointestinal disorders; although the
initial effects may escape notice for a day or
two, the presence of thallium salts can be
readily detected post-mortem. Until the 1960s,
the dominant use of thallium, as thallium
sulphate, was to poison rodents and ants. By
1972, when one quarter of all sick or dead bald
eagles had suffered thallium poisoning, the
United States banned its use as a poison,
followed in subsequent years by several other
countries. Thallium is used today in low-melting
glasses, photoelectric cells, switches, mercury
alloys for low-range glass thermometers
and thallium salts. William Crookes and
Claude-Auguste Lamy discovered thallium
independently in 1861, in residues of selenium-
bearing ores used in sulphuric acid production.
Both used newly developed flame spectroscopy,
in which thallium produces a prominent green
spectral line. Expecting to isolate tellurium after
removing selenium from the by-products, the
researchers instead found a new element.

30-SECOND TEXT
Jeffrey Owen Moran

*Traditionally, the
weapon of choice
for rat catchers, today
thallium – amongst
other uses – adds colour
to artificial gems and is
used in hi-tech infrared
light systems.*

Tl

LEAD

the 30-second element

Lead is one of the dozen or so elements that were known to peoples in the ancient world. Although it is not as strong as other metals, its softness and low melting point made it easy for them to work, and they found many uses for it. Lead forms a thin layer of white carbonate when exposed to air. This prevents further corrosion, making it ideal for roofing, sarcophagi and drains. The Romans used lead mined in Britain across the empire in these ways. Its colourful ores – black, white and red – were also used in cosmetics. The Latin word for lead is *plumbum*, which explains its chemical symbol Pb and the origin of words such as 'plumbing'. Its high density gave lead further applications as weights for fishing, in plumb lines and as dice. The Romans even used the highly soluble acetate salt of lead to sweeten wine. In modern times, lead was used to make printer's type and for bullets. Although the poisonous nature of lead compounds was recognized even in ancient times, it was only in the 20th century that many of its modern uses – including in domestic plumbing, paint, glass, solder and pewter – were restricted.

3-SECOND STATE

Chemical symbol: Pb
Atomic number: 82
Named: From the Anglo-Saxon word *lead* for this long-known metal

3-MINUTE REACTION

One of the remaining uses of lead is in car batteries. Invented in 1859, the lead-acid battery is relatively cheap and can deliver the high current needed to start an engine. When the battery discharges, electrodes immersed in sulphuric acid are converted to lead sulfate. Charging reverses the process. If the lead sulphate rests for too long, it begins to form crystals, and recharging becomes difficult – this is why it is important to maintain the charge.

RELATED ELEMENTS
See also
TIN (Sn 50)
page 126

FLEROVIUM (Fl 114)
page 150

3-SECOND BIOGRAPHIES
GASTON PLANTÉ
1834–89
French physicist, inventor of the lead-acid battery

THOMAS MIDGLEY
1889–1944
American chemist and mechanical engineer who added tetraethyl lead to petrol

ANSELM KIEFER
1945–
German artist who uses lead in many of his works

30-SECOND TEXT
Hugh Aldersey-Williams

Durable, highly resistant to corrosion, widely used by the Romans – lead has had many applications over the centuries. Alchemists saw lead as the oldest metal.

dioxygen Another name for diatomic oxygen, O_2, which forms one fifth of the earth's atmosphere. Known in common usage just as 'oxygen', O_2 is in fact one of several allotropes of oxygen: others include atomic oxygen (O_1), ozone (O_3) and tetraoxygen (O_4).

electronegativity The power of an atom in a molecule to attract electrons. The Pauling scale of electronegativity (named after American chemist Linus Pauling, who created the first electronegativity scale) stretches from 0.7 (francium) to 3.98 (fluorine).

graphite Allotrope of carbon, made in the earth's crust from the remains of plants. Very soft and a dark grey-to-black colour, graphite can be used to make a dark mark and was named from the Greek *graphein* ('to write/draw'). It is graphite and not lead that forms the pencil lead used for writing.

hydrogen atom The simplest of all the elements, the hydrogen atom has a single negatively charged electron orbiting a nucleus containing a single positively charged proton. Under normal conditions hydrogen is a diatomic gas, formula H_2.

hydrogen bond Interaction between a hydrogen atom and an electronegative atom (typically of oxygen, fluorine or nitrogen) in another molecule. Hydrogen bonds form between water molecules. Strictly speaking a hydrogen bond is not a chemical bond but an electromagnetic attraction between a partial positive charge on part of one molecule (in water, a hydrogen atom) and a partial negative charge on part of another molecule (in water, an oxygen atom). Each water molecule has hydrogen bonds to four others. The attraction of the hydrogen bonds gives water a high boiling point (100°C/212°F).

island of stability Theoretical group of isotopes of transuranic elements predicted to be much more stable than those currently known. These isotopes would have a longer half-life, making them easier to study.

mantle Layer of the earth, some 2900 km (1800 miles) thick, between the surface (crust) and the outer part of the planet's core. Diamonds are formed in the mantle around 150 km (90 miles) below the earth's surface, mostly from carbon that has been within the earth since its formation, at temperatures of at least 1000°C (about 2000°F). The diamonds arrive at or near the surface as a result of deep volcanic activity.

ozone An allotrope of oxygen, ozone is a triatomic molecule containing three oxygen atoms, O_3. It is found in the earth's upper atmosphere, where it screens us from harmful ultraviolet radiation from the sun.

paramagnetic Attracted by a magnetic field, but not retaining magnetic properties after the field is removed. By contrast, ferromagnetic materials form permanent magnets when in a magnetic field; they remain magnetic after the magnetic field is taken away.

triatomic molecule A molecule containing three atoms.

vulcanization Chemical process that strengthens natural or synthetic rubber, making it more resistant and improving its elasticity. Developed in 1839 by American inventor Charles Goodyear (1800–60), it uses sulphur to make cross links between carbon polymers in natural rubber. The process is named after Vulcan, the ancient Roman god of fire. Tyres, hockey pucks and hoses are among the products made from vulcanized rubber. The process is performed at temperatures of 140–180°C (280–360°F).

NON-METALS

Of the elements generally classified as non-metals, hydrogen is in group 1 of the periodic table, and the others are in groups 13–18. The non-metals are not good conductors of heat or electricity. At room temperature they are typically gases (such as hydrogen or oxygen) or solids (for example, carbon).

Non-metals

	Symbol	Atomic Number
Hydrogen	H	1
Carbon	C	6
Nitrogen	N	7
Oxygen	O	8
Phosphorus	P	15
Sulphur	S	16
Selenium	Se	34
Flerovium	Fl	114
Ununseptium	Uus	117

HYDROGEN

the 30-second element

The most abundant and simplest of the elements – a single proton and an electron – hydrogen has been around since atoms coalesced following the Big Bang more than 13 billion years ago. Even though stars constantly eat into that initial production, hydrogen still forms 75 per cent of the detectable content of the universe. This light, colourless, highly flammable gas is essential for life: without it, we wouldn't have the sun's heat, water or the organic compounds that form the building blocks of living things. It is only because hydrogen forms weak bonds between molecules that water is liquid on the earth. Without these 'hydrogen bonds', it would boil below –70°C (–94°F). Hydrogen was isolated by British scientist Henry Cavendish in the mid-18th century and soon put to use; its lightness makes it buoyant in air, a natural filling for balloons. French scientist Jacques Charles started this trend in 1783, but hydrogen really took off with dirigibles and zeppelins – until the destruction by fire of German airship LZ129 *Hindenburg* in 1937 doomed the hydrogen airship. More recently, hydrogen has been proposed as a replacement for fossil fuels in cars – when burnt, it produces no carbon dioxide, just harmless water.

3-SECOND STATE
Chemical symbol: H
Atomic number: 1
Named: From French
hydrogène ('watermaker')

3-MINUTE REACTION
In the sun and other stars, hydrogen burns in a fusion reaction. Hydrogen nuclei (protons) combine to produce the next heaviest element, helium, releasing energy. Every second, 600 million tons of hydrogen fuse into helium in the sun, producing 400 billion billion megawatts of power. Hydrogen also acts as more conventional fuel in some National Aeronautics and Space Association (NASA) rockets, where its impressive power per unit weight ensures maximum thrust.

RELATED ELEMENTS
See also
HELIUM (He 2)
page 62

3-SECOND BIOGRAPHIES
PARACELSUS
1493–1541
German-Swiss doctor who discovered hydrogen but did not realize what it was

HENRY CAVENDISH
1731–1810
British scientist who first isolated hydrogen, calling it 'inflammable air'

ANTOINE LAVOISIER
1743–94
French chemist who named hydrogen in 1783

30-SECOND TEXT
Brian Clegg

Nuclear fusion of hydrogen atoms in the sun powers life on the earth, but dreams of using hydrogen to enable long-distance flight ended with the Hindenburg airship disaster of 1937.

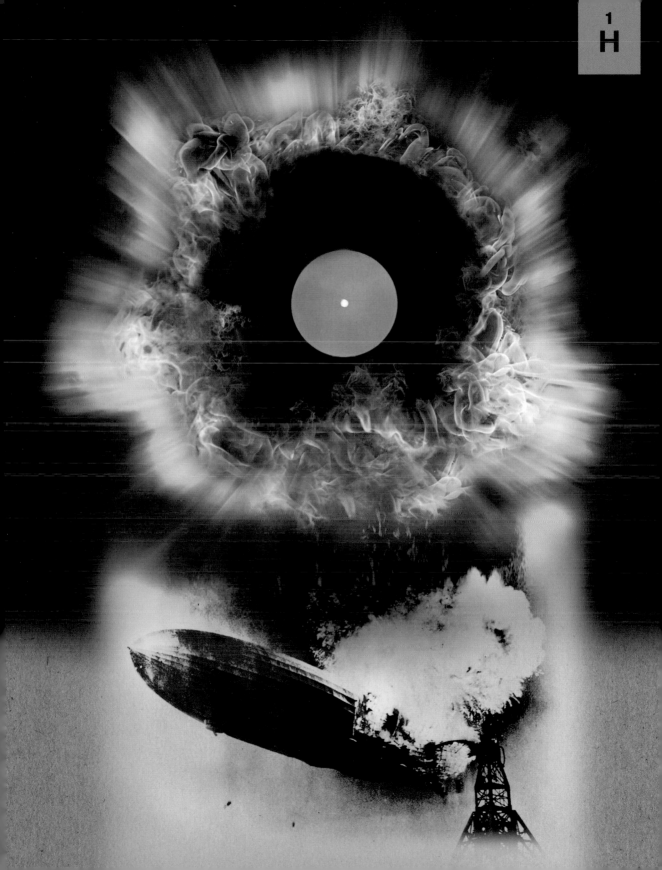

CARBON

the 30-second element

For mystique, kudos and significance, no element beats carbon. The readiness of carbon atoms to bond with each other into chains, rings and other complex frameworks enables them to provide the scaffolding of life's molecules. Carbon is the fabric of diamonds, cooked up deep in the earth's mantle mostly from carbon trapped inside the planet when it formed. Diamond's dirty sibling graphite is pure carbon too, made from sedimented remnants of dead plants that are squeezed and altered in the earth's crust, via coal, until only carbon remains. Diamond and graphite differ in how their carbon atoms are joined: to four neighbours in a three-dimensional network in diamond and to three neighbours in sheets of hexagonal rings in graphite; this makes the difference between superhard, glittering transparency and soft, metallic greyness, between diamond in drill bits and graphite in lubricants and pencils. Some old stars contain large amounts of carbon; their cores may contract to form planet-sized diamonds. But diamond and graphite aren't carbon's only elemental states. Individual sheets of graphite (graphene) conduct electricity and could have uses in electronics. These sheets can curl into tiny tubes (carbon nanotubes) and hollow molecular-scale shells (fullerenes) – all central to nanotechnology.

3-SECOND STATE
Chemical symbol: C
Atomic number: 6
Named: From Latin *carbo*, coal/charcoal

3-MINUTE REACTION
Carbon has two stable isotopes, carbon-12 and carbon-13. Carbon-14 (C-14) disintegrates spontaneously by radioactive decay, but it is constantly formed in the atmosphere by nuclear reactions of nitrogen sparked by cosmic rays. Some of this C-14 is woven into living organisms; when they die and stop replenishing it, the gradual decay of C-14 can be used (in the technique of radiocarbon dating) to deduce the organism's age.

RELATED ELEMENTS
See also
NITROGEN (N 7)
page 140

OXYGEN (O 8)
page 142

SILICON (Si 14)
page 100

3-SECOND BIOGRAPHIES
ROBERT CURL, HARRY KROTO, RICHARD SMALLEY
1933– , 1939– & 1943–2005
American, British, and American chemists, co-discoverers of C_{60}, or buckyballs, for which they were awarded the Nobel Prize

MILDRED DRESSELHAUS
1930–
American physicist, pioneer of carbon nanostructures

30-SECOND TEXT
Philip Ball

All that glitters is not gold ... it might be a diamond, carbon's valuable allotrope. Around 80 per cent of mined diamonds, not for jewellery, are used in industry.

NITROGEN

the 30-second element

3-SECOND STATE
Chemical symbol: N
Atomic number: 7
Named: From origin
(Greek root *gen-*) of nitre
(=saltpetre)

3-MINUTE REACTION
In the industrial Haber process, nitrogen gas is combined with hydrogen to make ammonia (NH_3); in the Ostwald process ammonia is combined with oxygen to make nitric acid (HNO_3). Ammonia and nitric acid are used in the production of explosives and fertilizers. Nitric acid also has a role in woodworking, where it is applied to 'age' wood, and in metal etching, where it is used in a solution with water and alcohol.

Elemental nitrogen constitutes

78 per cent of the earth's atmosphere. Nitrogen was first recognized as an element in the modern sense by French chemist Antoine Lavoisier in 1787. Both nitrogen's natural isotopes (nitrogen-14 and nitrogen-15) are stable; 14 radioactive isotopes have been found. Radioactive nitrogen-13 is made for use in PET scans (positron emission tomography scans, an imaging process that produces three-dimensional images of body processes). Liquid nitrogen at −196°C (−320°F) is used principally as a refrigerant, for purposes including the preservation of reproductive cells and biological samples. There are various oxides (combinations of nitrogen with oxygen): one atom of each (nitric oxide, NO) makes a muscle relaxant; with more nitrogen it makes 'laughing gas', used as an anaesthetic, but more oxygen produces a toxic gas, found in vehicle exhausts – one of the sources of acid rain. Nitrogen is essential for life, being part of DNA, of all proteins, and of various signalling molecules. Most of the nitrogen in nature is atmospheric, and it is converted by soil bacteria into nitrates, in which groups of nitrogen and oxygen atoms combine with other elements. Modern agriculture depends on nitrate fertilizer, made from the atmospheric gas by industrial processes.

RELATED ELEMENTS
See also
CARBON (C 6)
page 138

OXYGEN (O 8)
page 142

PHOSPHORUS (P 15)
page 144

3-SECOND BIOGRAPHIES
ANTOINE LAVOISIER
1743–94
French scientist, the first to recognize clearly the elemental status of nitrogen, oxygen and hydrogen

DANIEL RUTHERFORD
1749–1819
Scottish chemist and botanist, first to isolate nitrogen, which he called 'noxious air'

30-SECOND TEXT
P.J. Stewart

N_2O (nitrous oxide) is the sweet-tasting 'laughing gas' sometimes used to kill pain, but NO_2 (nitrogen dioxide) is no laughing matter – a major air pollutant from vehicular exhausts.

OXYGEN

the 30-second element

Oxygen is the third most abundant element in the universe, due to its stable 'doubly magic' nuclear structure. (The theory of magic nuclear numbers is based on study of nuclei that are especially stable because they contain a particular – 'magic' – number of protons and neutrons; when there is a magic number of both protons and neutrons the nucleus is 'doubly magic'.) The element was recognized in 1774 by French chemist Antoine Lavoisier, overturning the phlogiston theory of combustion (which postulated that things contained a fiery element – phlogiston – released when they burnt) and leading to our modern understanding of energy, mass and heat. Lavoisier named the element *oxygène*, from Greek words meaning 'begetter of acids' because he believed it to be present in all acids. Elemental oxygen is a highly reactive oxidizing agent. In meteorites, oxygen is always found chemically bonded, usually in silicate minerals. The earth's atmosphere was originally oxygen-free, but today one fifth of the atmosphere consists of diatomic gas molecules, O_2. These are a by-product of the biological process of photosynthesis: sunlight + carbon dioxide → glucose + oxygen. Therefore scientists examining planets outside our solar system look for the presence of O_2 as an indication of life.

3-SECOND STATE
Chemical symbol: O
Atomic number: 8
Named: From Greek *oxys* ('acid' or 'sharp') and genes ('begetter')

3-MINUTE REACTION
Oxygen is an electronegative element with a valency of 2. The important O-H bond, found in water, alcohols, sugars and most acids, is associated with hydrogen bonding, an attraction between molecules that causes high boiling point and solubility in water. Liquid oxygen, boiling point –369°C (–632°F), is pale blue in colour and is strongly paramagnetic; it can be suspended between the poles of a strong horseshoe magnet.

RELATED ELEMENTS
See also
NITROGEN (N 7)
page 140

SULPHUR (S 16)
page 146

3-SECOND BIOGRAPHY
JOSEPH PRIESTLEY & CARL WILHELM SCHEELE
1733–1804 & 1742–86
German-Swedish chemist and British natural philosopher/theologian, co-discoverers of oxygen in 1772–74. Scheele called it 'fire air' and Priestley, 'dephlogisticated air'

ANTOINE LAVOISIER
1743–94
French nobleman and chemist, who identified oxygen as an element and named it

30-SECOND TEXT
Mark Leach

No combustion without oxygenation: what happens when you create a vacuum and remove oxygen from the equation.

PHOSPHORUS

the 30-second element

In 1669, phosphorus became the first element to be discovered since ancient times, and in spectacular fashion: German alchemist Hennig Brandt extracted it from the residue of dozens of buckets of urine provided by his neighbours in Hamburg, and he was rewarded with a substance that glowed in the dark. He kept his procedure secret, but Anglo-Irish natural philosopher Robert Boyle cracked it, and soon less smelly methods were found. The pure element, which takes different forms – white, red, violet and black – is too reactive to exist in nature. White phosphorus is a deadly poison. Its use in early matches caused bone disorders and 'phossy jaw' (toothaches, gum swelling, jaw abscesses, and eventual brain damage) in workers who handled it; in modern matches, red phosphorus instead is used in the striking surface. White phosphorus is today used in incendiary bombs, flares and smoke grenades. The main source of phosphorus is phosphate-rich rock, in which phosphorus and oxygen form the phosphate ion, combining mainly with calcium. Phosphorus is essential to life, forming part of DNA, of cell membranes and of the molecules that transfer energy around cells. Bones and teeth consist mainly of calcium phosphate, present also in milk.

SECOND STATE
Chemical symbol: P
Atomic number: 15
Named: From Greek
for the Morning Star,
Venus: *phoros* ('bearer')
of *phos* ('light')

3-MINUTE REACTION
Phosphorus figures in many important compounds. In bone ash, it is an ingredient of porcelain. A sodium compound is used as a water softener and in laundry powders. Joined to organic molecules, phosphorus is found in flame retardants, herbicides and insecticides, and also in the now outlawed nerve poison sarin, which was developed in Germany and used in the 20 March 1995, gas attack on the underground railway in Tokyo, Japan, by the Aum Shinrikyo cult.

RELATED ELEMENTS
See also
NITROGEN (N 7)
page 140

SULPHUR (S 16)
page 146

ARSENIC (As 33)
page 104

3-SECOND BIOGRAPHIES
ROBERT BOYLE
1627–91
Anglo-Irish natural philosopher who improved Brandt's method of producing phosphorus

HENNIG BRANDT
1630–1710
German alchemist who discovered phosphorus in his quest for the mythical Philosopher's Stone

30-SECOND TEXT
P.J. Stewart

Hellish element? The Greek words that give phosphorus its name translate in Latin to Lucifer, applied in Judeo-Christian tradition to the chief of the fallen angels, or Satan.

SULPHUR

the 30-second element

If the devil had an element, this would be it. Sulphur burns with pungent vigour, for which reason it was called brimstone (*byrnstan*, 'burning stone') in Old English. As such, according to the biblical Book of Genesis, it rained down on the sinful cities of Sodom and Gomorrah, while the Book of Revelation declares that the devil will join sinners in a lake of burning sulphur on Judgement Day. Flaming sulphur has caused plenty of suffering already: it was a component of the incendiary weapon 'Greek fire' (developed c. 672 BCE) and of gunpowder. Combined with oxygen, sulphur forms acrid oxides that dissolve in water to form sulphuric acid. On the planet Venus, sulphuric acid condenses into clouds that rain acid on to the planet's surface. Earth's acid rain is less potent, but corrodes stone and kills trees – this dilute acid forms from sulphurous fumes released when coal is burned. Despite all this, sulphur is essential for life. It is a component of two of the amino acids that make up proteins, and we each contain about 140 g (5 oz) of it. One theory proposes that reactions of sulphur at deep-sea vents initiated the origin of life.

RELATED ELEMENTS
See also
OXYGEN (O 8)
page 142

SELENIUM (Se 34)
page 135

3-SECOND BIOGRAPHIES
JABIR IBN HAYYAN
c. 721–c. 815
Also called 'Geber', Persian alchemist who proposed the theory that metals are generated by sulphur and mercury

CHARLES GOODYEAR
1800–60
American inventor who invented rubber vulcanization

30-SECOND TEXT
Philip Ball

Airborne fire ... On Venus, you would have to cope with sulphuric rain; Byzantine Greeks are believed to have combined sulphur with nitre (potassium nitrate) in 'Greek fire'.

3-SECOND STATE
Chemical symbol: S
Atomic number: 16
Named: From Arabic *sufra*, for the mineral form

3-MINUTE REACTION
Bonded to carbon, sulphur atoms act as links that bridge the chain-like molecules of certain carbon-based polymers. In hair and wool, two conjoined sulphur atoms bridge molecules of the main protein constituent keratin. Hair straightening uses a chemical to break these bonds and loosen the coils they induce. Sulphur bridges also link the carbon polymers in natural rubber during the stiffening process of vulcanization.

12 October 1941
Born in Hinnerjoki, near Turku, Finland

1961
Studies physics at the University of Turku, Finland

1967
Awarded a doctorate in physics, with significant research in quantum chemistry

1971
Publishes first paper on relativistic effects

1974–84
Appointed associate professor of quantum chemistry, Åbo Akademi, Finland

1984
Becomes professor of chemistry at the University of Helsinki

1995
Decorated by the President of Finland

2009
Awarded emeritus status at the University of Helsinki

2009–12
Chairman of the International Academy of Quantum Molecular Science (IAQMS)

2011
Pyykkö and colleagues propose that lead-acid batteries get most of their voltage from special relativistic effects

2012
Wins the Schrödinger Medal for his pioneering work in the field of quantum chemistry

PEKKA PYYKKO

'Half of chemistry is still undiscovered. We don't know what it looks like and that's the challenge', says Finnish theoretical chemist Pekka Pyykkö. Quantum chemists such as Pyykkö are using the theories of quantum mechanics and relativity, aided by developments in computer software, to push the boundaries of scientific research.

Born in 1941, Pyykkö studied at the University of Turku, and was awarded a doctorate in physics in 1967. One third of his PhD focused on quantum chemistry, which he had learnt in Uppsala, Sweden, and then proceeded to establish in Finland as an official field of study.

Pyykko became an associate professor in 1974 and professor of chemistry at the University of Helsinki in 1984, where he served until 2009. He has worked on relativistic quantum chemistry, augmenting quantum theory with Einstein's theory of relativity, and has made pioneering contributions to numerical quantum chemistry (the use of computers to solve the complex equations governing the theory).

In its current form the periodic table consists of seven periods, or rows; however, in 1969 Glenn T. Seaborg (see pages 44–45) proposed that an eighth period could exist. Pyykkö has used a computational model to plot 'periods eight and nine', and the 54 elements he believes they contain. If he's right, these will contain elements up to atomic number 172, which is far beyond any chemical element that has been synthesized so far (element 118). Each of the elements predicted in these new periods is estimated to have a very short half-life. However, it may be possible to manufacture the elements inside particle accelerators.

In 2012, he was awarded the Schrödinger Medal for his pioneering contributions to quantum chemistry. This field of study enables scientists to predict the properties of chemicals using fundamental laws of physics. It could offer some of science's most important advances – for instance, replacing drug testing on animals, developing the diagnosis and treatment of cancer, and supporting the search to find cleaner energy sources.

Pyykkö has predicted several completely new chemical species that have subsequently been discovered. These include the covalently bonded ion of $[AuXe]^+$, in which gold forms bonds with the extremely inert element of xenon. He has also predicted the existence of a triple bond involving atoms of gold with carbon and a new molecule of $[WAu_{12}]$ involving the metal tungsten (W) and gold.

FLEROVIUM

the 30-second element

Flerovium, officially named in 2012, is an artificial element, which exists only as a few fleeting atoms at a time. The possibility of elements heavier than uranium was imagined by French polymath Charles Janet in 1928, long before any were created. He provided places for elements up to number 120 in his periodic table, placing element 114 under carbon. The theory of magic nuclear numbers – which describes nuclei that are especially stable because they contain a particular number of protons and neutrons – suggested that 114 and 184 should be magic: an isotope of flerovium with 114 protons and 184 neutrons (flerovium-298) should be doubly magic and would lie on an 'island of stability', meaning that it and its neighbouring elements should be specially stable. More recently, however, it has been supposed that this point would be reached with element 120 or 126. Beginning in December 1998, physicists in Dubna, Russia, made a few atoms of element 114 by firing calcium-48 at plutonium targets. The early results were uncertain, but it is now accepted that in the course of several experiments isotopes 285 to 289 were produced; of these, the heaviest has the longest half-life, in one form more than a minute.

RELATED ELEMENTS
See also
PLUTONIUM (Pu 94)
page 46

UNUNSEPTIUM (Uus 117)
page 152

3-SECOND STATE
Chemical symbol: Fl
(formerly Uuq)
Atomic number: 114
Named: After the Flerov Laboratory of Nuclear Reactions in Russia, itself named after Russian physicist Georgy Flerov (1913–90)

3-MINUTE REACTION
The experiment at Dubna that synthesized flerovium used a particle accelerator in which calcium atoms were fired at 10 per cent of the speed of light towards plutonium targets. Initially, one single atom of flerovium-289 was made; later in the experiment – which ran for six months – two atoms of flerovium-288 were made. Before the element was officially named flerovium on 31 May 2012 it had the temporary name ununquadium (Uuq).

3-SECOND BIOGRAPHIES
CHARLES JANET
1849–1932
French polymath who devised the 'left-step' periodic table

GEORGY FLEROV
1913–90
Soviet nuclear physicist, initiator of the Soviet atomic bomb programme, whose name was given to flerovium

30-SECOND TEXT
P.J. Stewart

Because only a few atoms of flerovium have been made, chemists cannot be sure of its properties, but experiments suggest it is more volatile than expected, given its position in the periodic table.

UNUNSEPTIUM

the 30-second element

At the end of the 20th century, the periodic table was looking somewhat incomplete: The man-made heavier-than-uranium elements (the transuranics) that occupied the final row of actinide elements petered out after element 112, copernicium (Cn). However, the years since the end of the 20th century have delivered rapid progress in the search for new elements. Element 114, flerovium (Fl), was discovered in 1999; element 116 – originally ununhexium (Uuh), but now livermorium (Lv) – in 2000; element 118, ununoctium (Uuo), in 2002; elements 113, ununtrium (Uut) and 115, ununpentium (Uup), in 2004. The final row was completed with the discovery of element 117, ununseptium (Uus), in 2010 by a collaboration between Russian and American scientists in Dubna, Russia. Now the row is full and if more elements are discovered, it will be necessary to start a new one. Element 117, sometimes called 'eka-astatine', has the temporary name ununseptium. Although not known from experiment, because so little of the substance has been prepared, it is predicted that it will be a solid, halogen-like substance with a melting point in the range 340–550°C (644–1,022°F).

3-MINUTE REACTION
Transuranic elements are made by fusing atomic nuclei together. The short-lived products decay into more stable nuclei. It is by studying these daughter products that the original's nuclear structure is deduced and then corroborated. So far, only two isotopes of ununseptium – Uus-293 and Uus-294 – have been prepared (in vanishingly small amounts) with millisecond lifetimes. Theoretical studies suggest that Uus-326 and Uus-327 might survive for hundreds of years, or even longer than that.

RELATED ELEMENTS
See also
COPERNICIUM (Cn 112)
page 92

FLEROVIUM (Fl 114)
page 150

3-SECOND BIOGRAPHY
GLENN T. SEABORG
1912–99
American chemist, principal or co-discoverer of ten elements, who predicted the existence of other transuranics

30-SECOND TEXT
Mark Leach

Americans and Russians worked together in producing element 117. It was made by firing calcium ions at a berkelium target. It is the second-heaviest known element, after element 118.

NOTES ON CONTRIBUTORS

Hugh Aldersey-Williams is a writer and curator with interests ranging from science to architecture and design. He is the author of the best-selling *Periodic Tales: The Curious Lives of the Elements*. His latest book is *Anatomies: The Human Body, Its Parts and the Stories They Tell*. He has curated exhibitions at the Victoria and Albert Museum and the Wellcome Collection, and is presently working on an exhibition of artworks related to the chemical elements for Compton Verney.

Philip Ball is a freelance writer, and was an editor for *Nature* for more than 20 years. Trained as a chemist at the University of Oxford, and as a physicist at the University of Bristol, he writes regularly in the scientific and popular media, and has authored books including *H2O: A Biography of Water*, *Bright Earth: The Invention of Colour*, *The Music Instinct* and *Curiosity: How Science Became Interested in Everything*. His book *Critical Mass* won the 2005 Aventis Prize for Science Books. He has been awarded the American Chemical Society's Grady–Stack Award for interpreting chemistry to the public, and was the inaugural recipient of the Lagrange Prize for communicating complex science.

Brian Clegg is a science writer based in Wiltshire whose books include *The God Effect*, *Before the Big Bang*, *Inflight Science* and *Build Your Own Time Machine*. He has written for many newspapers and magazines from *The Wall Street Journal* and *Nature* to *Playboy* and *Good Housekeeping*, and is a regular contributor to the Royal Society of Chemistry's podcasts. Brian is a fellow of the Royal Society of Arts and edits the www.popularscience.co.uk book review site.

John Emsley was a lecturer and reader in chemistry at King's College London for 22 years and produced more than 100 original research papers. He was Imperial College's Science Writer in Residence from 1990 to 1997, during which time he also wrote for *The Independent*. From 1997 to 2002, he held a similar position in the Department of Chemistry at the University of Cambridge when he produced its newsletter Chem@Cam. His popular science books include *The Consumer's Good Chemical Guide*, winner of the Science Book Prize in 1995, *Nature's Building Blocks*, *The Shocking History of Phosphorus* and *The Elements of Murder*.

Mark Leach studied at Loughborough and Salford Universities and as an academic chemist has worked at universities in the UK and overseas. He is the owner of the chemistry web publisher Meta-Synthesis and is the author of *The Chemogenesis Web Book*, *The Chemical Thesaurus Reaction Chemistry*

Database and is the curator of the Internet Database of Periodic Tables. Mark is interested in the philosophy of chemistry and the philosophical role played by the periodic table.

Jeffrey Moran is a software developer, whose company, Electric Prism, specializes in web-based learning applications. His spiral design of the periodic table has been featured in the *New York Times* and is etched in granite on the floor of the entry portico at Mount Holyoke College's Kendade Hall Science Center. A metalworker from an early age, Jeff's interest in the elements developed initially from his experience with precious and industrial metals, and grew when he designed restoration materials for the facades of historic buildings in New York and other cities. He is the author of *History Atlas*, an interactive history project currently in development, and a former two-term supervisor of the Town of Woodstock, New York.

Eric Scerri received all his education in the UK. Following postdoctoral work at Caltech, he became a Lecturer in Chemistry and History & Philosophy of Science at the University of California at Los Angeles (UCLA), where he has been for the past 13 years. He is the author of several books on the periodic table and more than 150 research and magazine articles in chemistry, chemical education and

history, and the philosophy of science. He is also the founder and editor of the journal *Foundations of Chemistry*. Scerri frequently lectures around the world on the history and significance of the periodic table and appears on radio and television.

Andrea Sella studied chemistry in Toronto and Oxford, specializing in organometallic synthesis. He has published papers on catalysis and the dynamics of transition metal systems, structure and bonding in lanthanide complexes, and in materials synthesis. Since 2011, he has been Professor of Chemistry at University College London where, in addition to teaching and teaching development, he is very active in bringing chemistry to a wider public through public lectures, articles such as his Classic Kit column in *Chemistry World* and contributions to radio and television. He has also been developing a programme of sending university students into primary schools.

Philip Stewart is a retired lecturer in Plant Sciences from the University of Oxford, who has published on a variety of topics, including the effects of human culture on ecosystems. He produced *Chemical Galaxy: a New Vision of the Periodic System of the Elements*, aimed at exciting the interest of non-chemists and young people in chemistry.

RESOURCES

BOOKS

Periodic Tales: A Cultural History of the Elements, from Arsenic to Zinc
Hugh Aldersey-Williams
(Ecco, 2012)

Periodic Tales: The Curious Lives of the Elements
Hugh Aldersey-Williams
(Viking, 2011)

The Building Blocks of the Universe
Isaac Asimov
(Lancer Books, 1966)

Elegant Solutions
Philip Ball
(Royal Society of Chemistry, 2005)

The Elements: A Very Short Introduction
Philip Ball
(Oxford University Press, 2004)

Sorting the Elements
Ian Barber
(Rourke Publishing, 2008)

The Elements
P.A. Cox
(Oxford University Press, 1989)

Nature's Building Blocks
John Emsley
(Oxford University Press, 2001)

The Disappearing Spoon
Sam Kean
(Back Bay Books, 2011)

The Periodic Table
Primo Levi
(Everyman's Library, 1996)

The Periodic Table: Its Story and Its Significance
Eric R. Scerri
(Oxford University Press, 2007)

Selected Papers on the Periodic Table
Eric R. Scerri
(Imperial College Press, 2009)

The Periodic Table: A Very Short Introduction
Eric R. Scerri
(Oxford University Press, 2012)

A Tale of Seven Elements
Eric R. Scerri
(Oxford University Press, 2013)

MAGAZINES/JOURNALS

Education in Chemistry
www.rsc.org/education/eic/

Foundations of Chemistry
link.springer.com/journal/
volumesAndIssues/10698

Chemistry World
www.rsc.org/chemistryworld

WEBSITES

The Chemical Galaxy
www.chemicalgalaxy.co.uk
Exploration and interpretation of the
elements and the periodic table.

Chemistry in Its Element
www.rsc.org/chemistryworld/podcast/
element.asp
Collection of short podcasts on each
of the elements from the Royal Society
of Chemistry.

Eric Scerri
www.ericscerri.com

Hugh Aldersey-Williams
www.hughalderseywilliams.com

John Emsley's World of Chemistry
www.johnemsley.com

Periodic Spiral
www.periodicspiral.com
An interactive alternate periodic table,
developed in a spiral pattern and created
by software developer Jeff Moran.

Popular Science
www.popularscience.co.uk
Popular science book review site,
including a range of books on chemistry
and the elements.

Visual Elements
www.rsc.org/periodic-table
Interactive periodic table from the Royal
Society of Chemistry.

WebElements
www.webelements.com
Useful online source of information on the
elements and the periodic table, developed
by the University of Sheffield.

INDEX

ACKNOWLEDGEMENTS

PICTURE CREDITS
The publisher would like to thank the following individuals and organizations for their kind permission to reproduce the images in this book. Every effort has been made to acknowledge the pictures; however, we apologize if there are any unintentional omissions.

Corbis/Stefano Bianchetti: 58.
Corbis/Bettmann: 80.
Shutterstock/www.shutterstock.com.

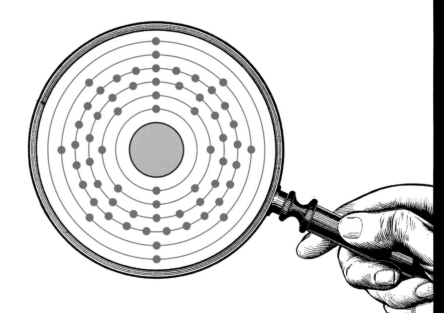